CLIMATE CHANGE AND ITS CAUSES, EFFECTS AND PREDICTION

CLIMATE CHANGE

LEGAL ISSUES AND CONTEXTS

CLIMATE CHANGE AND ITS CAUSES, EFFECTS AND PREDICTION

Additional books in this series can be found on Nova's website under the Series tab.

Additional E-books in this series can be found on Nova's website under the E-book tab.

LAWS AND LEGISLATION

Additional books in this series can be found on Nova's website under the Series tab.

Additional E-books in this series can be found on Nova's website under the E-book tab.

CLIMATE CHANGE AND ITS CAUSES, EFFECTS AND PREDICTION

CLIMATE CHANGE

LEGAL ISSUES AND CONTEXTS

BAILEY SAUNDERS
AND
ROSS A. DIAZ
EDITORS

New York

Copyright © 2013 by Nova Science Publishers, Inc.

All rights reserved. No part of this book may be reproduced, stored in a retrieval system or transmitted in any form or by any means: electronic, electrostatic, magnetic, tape, mechanical photocopying, recording or otherwise without the written permission of the Publisher.

For permission to use material from this book please contact us:
Telephone 631-231-7269; Fax 631-231-8175
Web Site: http://www.novapublishers.com

NOTICE TO THE READER

The Publisher has taken reasonable care in the preparation of this book, but makes no expressed or implied warranty of any kind and assumes no responsibility for any errors or omissions. No liability is assumed for incidental or consequential damages in connection with or arising out of information contained in this book. The Publisher shall not be liable for any special, consequential, or exemplary damages resulting, in whole or in part, from the readers' use of, or reliance upon, this material. Any parts of this book based on government reports are so indicated and copyright is claimed for those parts to the extent applicable to compilations of such works.

Independent verification should be sought for any data, advice or recommendations contained in this book. In addition, no responsibility is assumed by the publisher for any injury and/or damage to persons or property arising from any methods, products, instructions, ideas or otherwise contained in this publication.

This publication is designed to provide accurate and authoritative information with regard to the subject matter covered herein. It is sold with the clear understanding that the Publisher is not engaged in rendering legal or any other professional services. If legal or any other expert assistance is required, the services of a competent person should be sought. FROM A DECLARATION OF PARTICIPANTS JOINTLY ADOPTED BY A COMMITTEE OF THE AMERICAN BAR ASSOCIATION AND A COMMITTEE OF PUBLISHERS.

Additional color graphics may be available in the e-book version of this book.

LIBRARY OF CONGRESS CATALOGING-IN-PUBLICATION DATA

ISBN: 978-1-62257-847-4

Published by Nova Science Publishers, Inc. † New York

CONTENTS

Preface   vii

Chapter 1   Climate Change and Existing Law: A Survey of Legal Issues Past, Present, and Future   1
Robert Meltz

Chapter 2   Common-Law Climate Change Litigation after *American Electric Power v. Connecticut*   45
Robert Meltz

Chapter 3   Federal Agency Actions Following the Supreme Court's Climate Change Decision in *Massachusetts v. EPA:* A Chronology   63
Robert Meltz

Chapter 4   Climate Change Litigation: A Survey   77
Robert Meltz

Index   129

PREFACE

This book surveys existing law for legal issues that have arisen, or may arise in the future, on climate change and the governments responses, thereto. At the threshold of many climate-change-related lawsuits are two barriers-- whether the plaintiff has standing to sue and whether the claim being made presents a political question. Both barriers have forced courts to apply amorphous standards in a new and complex context. Efforts to mitigate climate change--that is, reduce greenhouse gas (GHG) emissions--have spawned a host of legal issues.

Chapter 1 – This report surveys *existing* law for legal issues that have arisen, or may arise in the future, on account of climate change and government responses thereto.

At the threshold of many climate-change-related lawsuits are two barriers—whether the plaintiff has standing to sue and whether the claim being made presents a political question. Both barriers have forced courts to apply amorphous standards in a new and complex context.

Efforts to mitigate climate change—that is, reduce greenhouse gas (GHG) emissions—have spawned a host of legal issues. The Supreme Court resolved a big one in 2007—the Clean Air Act (CAA), it said, does authorize EPA to regulate GHG emissions. Quite recently, a host of issues raised by EPA's efforts to carry out that authority were resolved in the agency's favor by the D.C. Circuit. Another issue is whether EPA's "endangerment finding" for GHG emissions from new motor vehicles will compel EPA to move against GHG emissions under other CAA authorities. Still other mitigation issues are (1) the role of the Endangered Species Act in addressing climate change; (2) how climate change must be considered under the National Environmental Policy Act; (3) liability and other questions raised by carbon capture and sequestration; (4) constitutional constraints on land use regulation and state actions against climate change; and (5) whether the public trust doctrine applies to the atmosphere.

Liability for harms allegedly caused by climate change has raised another crop of legal issues. The Supreme Court decision that the CAA bars federal judges from imposing their own limits on GHG emissions from power plants has led observers to ask: Can plaintiffs alleging climate change harms still seek monetary damages, and are state law claims still allowed? The one ruling so far says no to both. Questions of insurance policy coverage are also likely to be litigated. Finally, the applicability of international law principles to climate change has yet to be resolved.

Water shortages thought to be induced by climate change likely will lead to litigation over the nature of water rights. Shortages have already prompted several lawsuits over whether cutbacks in water delivered from federal projects effect Fifth Amendment takings or breaches of contract.

Sea level rise and extreme precipitation linked to climate change raise questions as to (1) the effect of sea level rise on the beachfront owner's property line; (2) whether public beach access easements migrate with the landward movement of beaches; (3) design and operation of federal levees; and (4) government failure to take preventive measures against climate change harms.

Other adaptation responses to climate change raising legal issues, often property rights related, are beach armoring (seawalls, bulkheads, etc.), beach renourishment, and "retreat" measures. Retreat measures seek to move existing development away from areas likely to be affected by floods and sea level rise, and to discourage new development there.

Natural disasters to which climate change contributes may prompt questions as to whether response actions taken in an emergency are subject to relaxed requirements and, similarly, as to the rebuilding of structures destroyed by such disasters just as they were before.

Finally, immigration and refugee law appear not to cover persons forced to relocate beca

Chapter 2 – Congressional inaction on climate change has led concerned parties to explore other ways to address climate change—including lawsuits seeking to establish climate change impacts as a common law nuisance.

The prospects for these common law suits are limited, owing in part to the unsuitability of private litigation for dealing with global problems like climate change. Recently, the outlook for federal common-law suits seeking injunctive relief vis-a-vis climate change became particularly dim. On June 20, 2011, the

Supreme Court ruled in *American Electric Power Co., Inc. v. Connecticut* that given EPA's Clean Air Act authority over greenhouse gas (GHG) emissions—affirmed by the Court a few years ago—the federal common law of nuisance in the area of climate change is "displaced." Federal courts may not use federal common law to add their own judge-made GHG emission standards to those of EPA.

The displacement of federal common law by *American Electric Power* is only one of three threshold issues that have bedeviled lawsuits seeking to establish climate change as a common law nuisance. The standing inquiry requires a plaintiff in federal court to show actual or imminent injury caused by the defendant, and the likelihood that the injury will be redressed by the requested relief. Each of these factors can pose difficulties for the climate-change plaintiff. Similarly, the political question doctrine has led some courts to dismiss common-law climate change suits on the ground that the issue is better left with the political branches. Of course, where *American Electric Power* applies and the case must be dismissed on displacement grounds, standing and political question doctrine are now less important.

American Electric Power raises several questions. First, with federal common law displaced in the area of climate change, are state common law claims viable? Two threats to such claims are the possibility of preemption by the Clean Air Act (the sounder argument is against preemption), and the influence of the Supreme's Court's aversion to judge-made law in the climate change area so evident in *American Electric Power*. A second question is whether *American Electric Power* displaces climate-change-based federal common law actions when the remedy sought is monetary rather than injunctive. Finally, if Congress eliminates EPA authority over GHG emissions and is silent as to federal common law actions, does federal common law cease to be displaced so that such actions are again possible?

In addition to *American Electric Power*, there are two other active cases raising common law nuisance claims as to climate change—both involving coastal damage. In *Village of Kivalina v. ExxonMobil Corp.*, a coastal Eskimo village is suing energy companies alleging that their GHG emissions have contributed to shoreline erosion, requiring relocation of the village. In *Comer v. Murphy Oil*, Gulf coast landowners are suing energy and chemical companies asserting that their GHG emissions intensified Hurricane Katrina, adding to plaintiffs' property damage. Both of these cases raise the above-noted issue whether *American Electric Power* applies to actions seeking monetary damages.

A second common law theory recently has entered the fray. Since May 2011, either a suit or rulemaking petition has been filed in every state arguing that the respective state has a "public trust" duty to the atmosphere that requires it to address climate change. A suit has also been filed against the United States on the same ground.

Chapter 3 – On April 2, 2007, the Supreme Court rendered one of its most important environmental decisions. In *Massachusetts v. EPA*, the Court held 5-4 that greenhouse gases (GHGs), widely viewed as contributing to climate change, constitute "air pollutants" as that phrase is used in the Clean Air Act (CAA). As a result, said the Court, the U.S. Environmental Protection Agency (EPA) had improperly denied a petition seeking CAA regulation of GHG emissions from new motor vehicles by saying the agency lacked authority over such emissions.

This report offers a chronology of major federal agency actions, mainly by EPA, that involve GHGs or climate change and that occurred in the wake of *Massachusetts v. EPA*. Most of the listed actions trace directly or indirectly back to the decision. Examples include EPA's "endangerment finding" for GHG emissions from new motor vehicles; the agency's standards for GHG emissions from new motor vehicles; its interpretation of "pollutants subject to regulation," the CAA trigger for requiring best available control technology (BACT) for such pollutants in "prevention of significant deterioration" areas; its guidance for determining BACT for GHG emissions; the "tailoring rule" (limiting the stationary sources that initially will have to install BACT and obtain CAA Title V permits based on their GHG emissions); and settlements of litigation seeking to compel new source performance standards (NSPSs) for GHG emissions from electric power plants and petroleum refineries. A few agency actions were included solely because of their relevance to climate change and their post-*Massachusetts* occurrence—for example, EPA's responses to California's request for a waiver of CAA preemption allowing that state to set its own limits for GHG emissions from new motor vehicles, and EPA's monitoring rule for GHG emissions.

Chapter 4 – The scientific, economic, and political questions surrounding climate change have long been with us. This report focuses instead on a relative newcomer: the legal debate. Though the first court decision related to climate change appeared 19 years ago, such litigation has proliferated in just the past six. Representatives of some suing organizations and states acknowledge that a prime cause for this litigation surge was inaction by Congress and the executive branch during the George W. Bush Administration with regard to mandatory constraints on greenhouse gas (GHG) emissions.

The court cases, decided and pending, arise in eight contexts. The first is the Clean Air Act (CAA). In *Massachusetts v. EPA*, the Supreme Court held that as to mobile sources of emissions (cars, trucks), EPA has authority under the act to regulate greenhouse gas (GHG) emissions. This decision puts pressure on EPA to move forward as well with regulation of GHGs from stationary sources (power plants, factories).

Second, litigation under wildlife statutes, particularly the Endangered Species Act, raises the possibility that the impacts of climate change on wildlife may constrain private activities that emit GHGs.

Third, energy statutes have been invoked. It has been held, for example, that under the Energy Policy and Conservation Act, the United States must monetize the benefits of reduced carbon emissions as part of setting light-truck fuel economy standards.

Fourth, various statutes requiring federal government analysis and information dissemination— the National Environmental Policy Act (NEPA), Global Change Research Act, and Freedom of Information Act—have generated climate-change litigation. NEPA suits make up the most numerous subset of this category. Courts agree that if a plaintiff can establish standing, NEPA can be used to compel agency consideration of the climate change effects of its actions.

Fifth, common law tort theories such as nuisance have been invoked, not yet successfully, to force cutbacks in GHG emissions, or payment of damages. Several cases are on appeal.

Sixth are the preemption suits. These challenge state regulation of GHG emissions from motor vehicles as preempted by the federal corporate average fuel economy standards or federal authority over foreign policy. The two rulings thus far have rejected these challenges, but are on appeal. California's suit attacking EPA's denial of its request for a waiver of federal preemption under the Clean Air Act has now been stayed, pending EPA reconsideration of the denial.

Seventh, chiefly with respect to coal-fired power plants, are suits under state utilities laws.

And eighth, one case asks whether existing general liability insurance policies cover climatechange-related liability.

Finally, the report discusses international law aspects of a nation's contributions to climate change, and offers some overview comments.

In: Climate Change: Legal Issues and Contexts   ISBN: 978-1-62257-847-4
Editors: B. Saunders and R. A. Diaz   © 2013 Nova Science Publishers, Inc.

Chapter 1

CLIMATE CHANGE AND EXISTING LAW: A SURVEY OF LEGAL ISSUES PAST, PRESENT, AND FUTURE[*]

Robert Meltz

SUMMARY

This report surveys *existing* law for legal issues that have arisen, or may arise in the future, on account of climate change and government responses thereto.

At the threshold of many climate-change-related lawsuits are two barriers—whether the plaintiff has standing to sue and whether the claim being made presents a political question. Both barriers have forced courts to apply amorphous standards in a new and complex context.

Efforts to mitigate climate change—that is, reduce greenhouse gas (GHG) emissions—have spawned a host of legal issues. The Supreme Court resolved a big one in 2007—the Clean Air Act (CAA), it said, does authorize EPA to regulate GHG emissions. Quite recently, a host of issues raised by EPA's efforts to carry out that authority were resolved in the agency's favor by the D.C. Circuit. Another issue is whether EPA's "endangerment finding" for GHG emissions from new motor vehicles will compel EPA to move against GHG emissions under other CAA authorities. Still other mitigation issues are (1) the role of the Endangered

---
[*] This is an edited, reformatted and augmented version of Congressional Research Service, Publication No. R42613, dated July 2, 2012.

Species Act in addressing climate change; (2) how climate change must be considered under the National Environmental Policy Act; (3) liability and other questions raised by carbon capture and sequestration; (4) constitutional constraints on land use regulation and state actions against climate change; and (5) whether the public trust doctrine applies to the atmosphere.

Liability for harms allegedly caused by climate change has raised another crop of legal issues. The Supreme Court decision that the CAA bars federal judges from imposing their own limits on GHG emissions from power plants has led observers to ask: Can plaintiffs alleging climate change harms still seek monetary damages, and are state law claims still allowed? The one ruling so far says no to both. Questions of insurance policy coverage are also likely to be litigated. Finally, the applicability of international law principles to climate change has yet to be resolved.

Water shortages thought to be induced by climate change likely will lead to litigation over the nature of water rights. Shortages have already prompted several lawsuits over whether cutbacks in water delivered from federal projects effect Fifth Amendment takings or breaches of contract.

Sea level rise and extreme precipitation linked to climate change raise questions as to (1) the effect of sea level rise on the beachfront owner's property line; (2) whether public beach access easements migrate with the landward movement of beaches; (3) design and operation of federal levees; and (4) government failure to take preventive measures against climate change harms.

Other adaptation responses to climate change raising legal issues, often property rights related, are beach armoring (seawalls, bulkheads, etc.), beach renourishment, and "retreat" measures. Retreat measures seek to move existing development away from areas likely to be affected by floods and sea level rise, and to discourage new development there.

Natural disasters to which climate change contributes may prompt questions as to whether response actions taken in an emergency are subject to relaxed requirements and, similarly, as to the rebuilding of structures destroyed by such disasters just as they were before.

Finally, immigration and refugee law appear not to cover persons forced to relocate because of climate change impacts such as drought or sea level rise.

INTRODUCTION

This report surveys *existing* law for legal issues that have arisen, or may arise in the future, on account of climate change and government responses thereto. The reader interested in proposals for *new* laws to deal with climate change is referred to other works.[1] Of course, while this report covers many of the major legal issues that have emerged or may do so, the endless ramifications of climate change preclude any claim to exhaustiveness.

The report takes as its point of departure the current scientific consensus that climate change is occurring and, to the degree it continues, will cause sea level rise and extreme weather events.[2] Inclusion of some legal issues was based further on the predominant scientific view that human activities are contributing to climate change.[3]

Finally, it should be noted that the discussion of several topics in this report likely would have to be substantially modified, or possibly deleted, if Congress were to enact comprehensive climate change legislation. Such legislation might limit or displace the role of certain existing statutes— the Clean Air Act and the Endangered Species Act being prime candidates—or common law in addressing climate change.

I. THRESHOLD BARRIERS TO LITIGATION

Federal courts have evolved a variety of gatekeeper doctrines to ensure that only certain plaintiffs and certain types of claims can invoke their jurisdiction. Two of these doctrines, standing and political question, have posed daunting barriers for plaintiffs in climate change cases.

Standing doctrine. This principle flows from Article III of the Constitution, which limits the jurisdiction of courts created under that article (such as federal district courts) to "cases" or "controversies." These words are construed to require a person who sues in an Article III court to show (1) "injury in fact" (existing or imminent), (2) "causation" (described as a fairly traceable connection between the injury in fact and the defendant's conduct), and (3) "redressability" (meaning that plaintiff's injury is likely to be remedied by the relief plaintiff seeks).[4] A plaintiff not satisfying any of these elements is said to lack standing; his or her suit will be dismissed.

It should be apparent that a plaintiff complaining of injury from climate change may be thwarted by any of the three standing requirements. For example, how does such a plaintiff show the second element, causation? How does he show, say, that a drought that destroyed his crops was caused by climate change—indeed, by climate change to which the defendant's greenhouse gas (GHG) emissions contributed?[5] To be sure, in two climate change decisions, *Massachusetts v. EPA* in the Supreme Court[6] and *American Elec. Power Co. v. Connecticut* in the Second Circuit,[7] Article III standing was found—but specifically for *state* plaintiffs.[8] *Massachusetts* asserted that states are entitled to "special solicitude" when seeking to establish standing,[9] and both decisions noted the sovereign status of states as *parens patriae* (literally, father of the country).[10] Case law since these decisions, however, has rejected their extension to private plaintiffs, who have often encountered difficulty establishing standing in climate change cases.[11] True, such plaintiffs may seek to avoid Article III standing issues by attempting to establish standing in state courts. But if, as is likely, the lawsuit takes aim at GHG emissions from out-of-state sources, the defendants are likely to remove the case to federal court under federal question or diversity jurisdiction. Thus the question of Article III standing likely will need to be faced.

A specialized issue is whether Indian tribes, by virtue of their inherent sovereignty, should also be able to establish standing through *parens patriae* status.[12] The argument for tribal *parens patriae* standing was rejected by the district court in *Native Village of Kivalina v. ExxonMobil Corp.*, a case in which an Eskimo village seeks damages for coastal erosion allegedly caused by climate change to which the defendants' GHG emissions assertedly contribute.[13] The case is now on appeal to the Ninth Circuit.

Political question doctrine. While standing asks whether there is a proper *plaintiff* before the court, political question doctrine asks whether there is a justiciable *claim*. The doctrine seeks to restrain courts from inappropriate interference in the business of the other branches of government—often because resolving the issue necessarily involves policy determinations. Six factors indicating a non-justiciable political question (any one of which may be dispositive) were famously stated by the Supreme Court in *Baker v. Carr* in 1962.[14] Of these, the first three have played a role in the climate-change nuisance cases:

> a textually demonstrable constitutional commitment of the issue to a coordinate political department; or a lack of judicially discoverable and manageable standards for resolving it; or the impossibility of deciding [the issue] without an initial policy determination of a kind clearly for nonjudicial discretion....

Baker made clear it was setting a high threshold for nonjusticiability; since it was decided a half-century ago, the Court has found few issues to present political questions. But the doctrine has been ubiquitous in the nuisance-based climate change litigation with more courts rejecting such claims on that ground than not.[15]

Addendum. At this point, the reader is referred to Section III.A., "A. Liability After *American Electric Power Co., Inc. v. Connecticut*," which discusses yet another litigation barrier: federal displacement of common-law-based climate change claims by the Clean Air Act. This barrier, where it applies, makes it unnecessary for courts to reach the standing and political question issues in the case, and thus allows them to avoid the abstruse questions raised by those defenses.

II. MITIGATION—REDUCING GHG EMISSIONS

Proactive responses to climate change are usually grouped under one of two headings: mitigation and adaptation. This section treats some of the legal issues raised by mitigation. Sections IV and V compile some of the legal issues associated with adaptation.

A. *Massachusetts v. EPA* and EPA's GHG Rules under the Clean Air Act

In 2007, the Supreme Court answered a fundamental Clean Air Act (CAA) question. The act, it found in *Massachusetts v. EPA*,[16] gives EPA authority to regulate GHG emissions. Such authority is granted, said the Court, because the CAA term "air pollutant" is defined sufficiently broadly in the act to include GHGs. Moreover, the Court added, the CAA forecloses an EPA decision not to regulate GHGs or any other air pollutant simply because the administration in power may have policy qualms—for example, due to a preference for non-regulatory approaches. In light of these determinations, the Court instructed EPA to reconsider its 2003 denial of a petition asking it to regulate GHG emissions from new motor vehicles, a denial EPA had based on the Court-rejected reasons.

Following this seminal decision, EPA set about the task of adapting the CAA to address climate change. In doing so, the agency confronted a statute more comfortably suited to regional air pollution problems, the opposite of climate change with its global nature. Four EPA actions in that effort are, in chronological order—

- The "timing rule." This "rule" is actually an EPA memorandum. 73 Fed. Reg. 80,300 (2008). The memorandum narrowly interprets the CAA phrase "pollutant subject to regulation under this act"[17] to include only pollutants regulated by *actual, not potential future*, emission limits. To explain, in Prevention of Significant Deterioration (PSD) areas—areas that are cleaner than national standards require—the CAA requires only a "pollutant subject to regulation under this act" to be controlled by potentially expensive "best available control technology" (BACT).[18] Since there were no "actual" GHG regulations under the CAA when the memorandum was issued, this meant that for a while at least, new major emitting facilities in PSD areas did not have to install BACT for GHG emissions. In 2010, EPA finalized a reconsideration of this rule and made clear that even under its "actual, not potential future" view, PSD requirements would kick in on January 2, 2011, when the "tailpipe rule" (below) took effect. 75 Fed. Reg. 17,004 (2010).
- The "endangerment finding," in which EPA determined that GHG emissions from new motor vehicles "cause, or contribute to, air pollution which may reasonably be anticipated to endanger public health or welfare," per CAA Section 202(a)(1).19 74 Fed. Reg. 66,496 (2009). This was pursuant to the reconsideration of the Section 202 petition ordered in Massachusetts. The finding has no effect on outside parties in itself; its importance is that it triggers a duty under CAA Section 202(a) for EPA to promulgate emission standards for new motor vehicles—see immediately below.
- The "tailpipe rule," in which EPA and the National Highway traffic Safety Administration jointly set GHG emission standards and fuel economy standards for 2012-2016 model year light-duty vehicles, 75 Fed. Reg. 25,323 (2010).
- The "tailoring rule," 75 Fed. Reg. 31,514 (2010). This rule is to relieve the overwhelming permitting burdens that EPA asserts would, in the absence of the rule, fall on PSD and Title V permitting authorities beginning January 2, 2011, when EPA's tailpipe rule took effect. When that happened, the PSD part of the CAA requires by its terms that PSD permits be issued (and BACT applied) for every new major emitting facility in the PSD area that emits more than either 100 or 250 tons of GHGs annually, depending on the source. This is a huge number of sources, so the tailoring rule sets much higher tonnage thresholds for 2011, gradually diminishing, EPA hopes, in

following years. That way, EPA expects, federal and state permitters will have time to develop routines for processing the extremely large number of permit applications.

Ninety-five petitions for review challenging these EPA actions, plus EPA's historic interpretation of the PSD section of the CAA, were filed in the D.C. Circuit. A marathon two days of oral argument before the court ensued. On June 26, 2012, the court handed EPA a resounding victory, unanimously upholding in firm language all four of EPA's actions.[20] Briefly, the court held that EPA's endangerment finding is adequately supported by the administrative record. The tailpipe rule, it said, is supported by the CAA's plain text and need not consider the rule's consequences for stationary sources of emissions. Turning to the stationary source regulations—the timing rule and tailoring rule—the court first found the PSD portion of the act to cover GHGs. The court then held it could not reach the merits of the rule challenges because petitioners lacked standing in light of their failure to show injury from the rules. For example, the tailoring rule produced a benefit for petitioners, not an injury, since without the rule an even *greater* number of sources would be subject to PSD and Title V permitting.

Assuming, as most observers do, that reversal on further review is unlikely, the D.C. Circuit ruling means at least two things. First, other adjudicative and administrative efforts involving GHG emissions regulation can now proceed. For example, with the endangerment finding under Section 202 now upheld, any EPA endangerment finding under Section 111,[21] governing new source performance standards for stationary sources, will be on firmer ground. That removes a stumbling block to EPA development of new source performance standards for GHG emissions from stationary source categories, such as those currently being finalized under court settlement for fossil-fuel-fired power plants and petroleum refineries.[22] Second, with the judicial option likely closed, states and industries opposed to EPA's efforts to address climate change through the CAA have few options left other than pressing Congress to curtail or eliminate EPA's CAA authority to deal with GHG emissions.

B. Legal Consequences of EPA's Endangerment Finding

With EPA's endangerment finding for new motor vehicle GHG emissions likely having survived judicial challenge, one question comes to the fore: does the finding, made under CAA Section 202, legally compel the agency to make endangerment findings for GHG emissions under other sections of the act that use similar endangerment language for other emission sources? Such subsequent endangerment findings would require, or at least authorize, EPA to regulate GHG emissions under those sections. CRS has explored this question in a separate report.[23] Briefly, that report concludes as follows:

First, the CAA section most likely to require EPA regulatory action after the Section 202 endangerment finding is Section 111. Section 111 requires EPA to set performance standards for those categories of new stationary sources of emissions that "cause, or contribute significantly to, air pollution which may reasonably be anticipated to endanger public health or welfare." The word "significantly," not present in Section 202, suggests that any legal compulsion created by the Section 202 endangerment finding might be limited to those new-source categories with the most prodigious GHG emissions. Section 111, however, affords EPA wide discretion in setting new source performance standards. As Section II.A. notes, EPA has already moved to use Section 111 against GHG emissions, pursuant to litigation settlements.

Second, two other CAA provisions that might be triggered by the Section 202 endangerment finding are Section 108,[24] requiring national ambient air quality standards, and Section 115,[25] which requires states to revise their implementation plans to prevent or eliminate the endangerment of public health or welfare in a foreign country. As to these sections, however, the arguable infeasibility of achieving the regulatory goals—even if GHG emissions in the United States are significantly reduced, atmospheric concentrations would decline little—may give EPA room to argue that regulatory action is not mandatory. Other endangerment-triggered sections of the CAA can be distinguished from Section 202(a) by their explicit terms, and so likely would not be triggered by the 202(a) endangerment finding—or at least do not impose on EPA a *mandatory* duty to promulgate GHG emission limits.

C. Use of the Endangered Species Act to Restrict GHG Emissions[26]

Some cast the Endangered Species Act (ESA) as a tool aggressive environmental groups may use to thwart projects that produce GHGs. Under this view, plaintiffs would claim that a project's GHG emissions, by contributing to climate change that brings about adverse habitat change, are causing a "take" of protected species in violation of the ESA.[27] For example, a suit could claim that any project that contributes to warmer seas harms, hence "takes," certain listed coral species. However, no case law can be found on this legal argument, either accepting or rejecting it.

Instead of alleging takes of species, lawsuits connecting the ESA to climate change typically are based on how an agency considered climate change when making other determinations: listing a species;[28] designating critical habitat;[29] or issuing a Biological Opinion.[30] The ESA requires that the Fish & Wildlife Service (FWS) consider the effects on habitat, at least in part, for all of those determinations.[31]

Accordingly, climate change evaluations long have been part of ESA decision-making, but only to the extent that the climate's effects on habitat are linked to a species.

Case law does not show that the ESA is used as an enforcement tool to make climate change arguments. In the handful of cases where ESA challenges were directed at federal projects related to power plants, only one involved climate change allegations, *Palm Beach County Environmental Coalition v. Florida*, and it was not clear whether those claims were premised on the ESA or on another legal basis.[32]

Despite the apparent lack of litigation premised on climate change *taking* species, some regulatory changes were made to limit lawsuits based on that cause of action.

In 2008, FWS changed the regulations that dictated how a service considered impacts of federal projects on listed species.[33] Those regulations were effective only from January 15, 2008, to May 5, 2008, after Congress acted to halt them in P.L. 111-8.[34]

During that period of regulatory change, definitions related to the effects of an agency action were modified to "reinforce the Services' current view that there is no requirement to consult on [greenhouse gas] emissions' contribution to global warming and its associated impacts on listed species."[35] Despite the revocation of those changes, it does not appear that the scope of effects has expanded, likely due to the fact that the regulations already limited review to those effects with a reasonable certainty to occur.[36]

Another regulatory change of the same time period is still in place. It restricts lawsuits claiming incidental takes of polar bears to instances where the agency action occurs in the state of Alaska.[37]

D. Government Restrictions on Private Activities That Generate GHGs or Reduce Carbon Sinks as Possible Takings of Private Property

Government restrictions on the use of private land always raise the prospect of landowners filing regulatory takings claims under the Fifth Amendment Takings Clause,[38] if such restrictions eliminate much of the land's value. Thus, government prohibition of, say, building a coal-fired powerplant on GHG-emitting grounds may generate a takings challenge if the proposed project site is substantially devalued. Research fails to reveal any court decisions in this category, but it can be said that regulatory takings claims in general are rarely successful, usually because other economic use of the site can be made.

Development restrictions on privately owned forests and wetlands on the basis of their carbon-sink value may also give rise to takings claims. A carbon sink is a natural or artificial reservoir that stores some carbon-containing compound. While the oceans are by far the largest carbon sink, in the form of dissolved carbon dioxide, forests and wetlands are significant repositories. The "public interest review" conducted by the Corps of Engineers when applications are submitted for wetlands development[39] would seem sufficiently broad to allow Corps consideration of a wetland's carbon-sink value. Again, however, research fails to reveal any court decisions as yet. Historically, though, takings challenges to development prohibitions in wetlands have shown a better chance of success than with development prohibitions generally, because a development-barred wetland may have no economic use whatsoever.

E. Consideration of Climate Change in Environmental Impact Statements

It is no longer in doubt that the National Environmental Policy Act (NEPA)[40] requires a federal agency to consider climate change impacts—those the agency's proposed project may contribute to, and those affecting the proposed project—in environmental impact statements (EISs).[41] The very first appearance of climate change in a reported court decision was in a NEPA case,[42] and the numerous NEPA/climate-change decisions since have never doubted that where sufficiently serious and causally connected to the project, climate change impacts should be discussed.[43] Draft guidance from the Council on Environmental Quality (CEQ) also makes the point.[44]

Still, clear thresholds triggering EIS inclusion have yet to emerge from the court decisions. CEQ suggests in its draft guidance that when federal activity is subject to GHG emissions accounting requirements, such as CAA reporting requirements that apply to stationary sources that directly emit 25,000 metric tons or more of CO_2-equivalent GHG on an annual basis, the agency should include this information in the NEPA documentation for consideration by decision makers and the public. CEQ expressly disclaims, however, that it intends 25,000 metric tons per year as the emission level that constitutes a "major federal action significantly affecting the quality of the human environment,"[45] NEPA's trigger for requiring an agency to prepare an EIS.

In addition to the federal NEPA, many states have NEPA-like statutes for evaluating proposals of state agencies. A full review of the legal issues raised by climate change under these "little NEPAs" is beyond the scope of this report. An example is the split in the California courts on whether projected future conditions (as in a climate-changed world) rather than current ones can be used as the baseline for evaluating the environmental impacts of proposed state projects.[46]

F. Carbon Capture and Sequestration[47]

While most proposals to mitigate climate change have focused on limiting GHG emissions, a prominent mitigation alternative is carbon capture and sequestration (CCS). CCS is a process whereby CO_2 emissions would be "captured" at their source and then stored or "sequestered" either underground or elsewhere, rather than being released into the atmosphere. Frequently, this storage/sequestration would take place underground.

Large-scale CCS technology is still in the early stages of development. Therefore, there are a number of operational questions to be answered before we can fully understand all the legal issues that may arise. However, because the development of CCS technology could well depend in part upon the resolution of some of these legal issues, it is important to understand them as the CCS debate continues. Among the emerging legal issues associated with CCS technology are (1) who owns and controls the underground pore space where the CO_2 would be "sequestered" under many of the CCS facility concepts proposed, in particular is pore space part of the surface estate or mineral rights under traditional property law principles; (2) which federal and state agencies would permit and regulate CO_2 pipelines transporting the gas from the point of generation to the sequestration site under the existing framework for pipeline regulation; and (3) concerns over liability exposure that may hinder development of CCS technology.[48]

G. Constitutional Barriers to State Action

Two federal constitutional constraints on state action, preemption and the dormant commerce clause, have played a role in blocking state efforts to restrict GHG emissions.

1. Preemption

Two federal statutes have been invoked to argue for federal preemption of state laws affecting GHG emissions: the CAA and the Energy Policy and Conservation Act (EPCA). The CAA, while not generally preempting state regulation of stationary source emissions, does preempt state standards "relating to" the control of emissions from new motor vehicles.[49] An exception is that EPA may waive CAA preemption for vehicle emission standards in California, should that state so request,[50] whereupon states with standards identical to California's also participate in the waiver.[51] EPCA, for its part, is not directly concerned with emissions. Rather, it authorizes federal promulgation of corporate average fuel economy standards ("CAFE standards"),[52] then dictates that when a CAFE standard is in effect, a state may not regulate in a manner "related to" such fuel economy standards.[53] No California waiver or other waiver is authorized.

An obvious ambiguity exists as to when a state action is "relating to" or "related to" the relevant federal action, and thus preempted. For example, one case dealt with city regulations reducing the rates at which taxicab owners could lease vehicles to drivers if the vehicle did not have a hybrid engine. The court found it "likely" (the standard for obtaining a preliminary injunction) that the regulations effectively required cab owners to buy only hybrid vehicles, so that the regulations were "relating to" the control of emissions under the CAA and "related to" CAFE standards under EPCA. So finding, the court held that plaintiffs had shown a likelihood of success in showing preemption, and a preliminary injunction was granted.[54]

It is also unclear at what point a state's actions restricting GHG emissions are preempted as interfering with national foreign policy, given the long history of U.S. involvement in international negotiations over GHG emissions.[55] The issue has been raised in litigation.[56]

2. Dormant Commerce Clause

Quite recently, in *Rocky Mountain Farmers Union v. Goldstene*, a federal district court ruled that California's Low Carbon Fuel Standard (LCFS) offends the Constitution's "dormant commerce clause."[57] The dormant commerce clause, a judicially created corollary of the Constitution's Commerce Clause,[58] bars a state from discriminating against commerce based on its out-of-state origin, and, even in the absence of discrimination, bars a state from imposing "undue burdens" on interstate commerce. Here, the court found that the LCFS discriminated against out-of-state corn-derived ethanol while favoring in-state corn ethanol, and impermissibly regulated extraterritorial conduct. In addition, said the court, the state had failed to show a lack of alternative, nondiscriminatory ways to reduce GHG emissions. The LCFS regulations are a part of California's attempts, under a state enactment, to reduce GHG emissions in California to 1990 levels by 2020.[59]

An intriguing question is whether *Rocky Mountain Farmers Union* may lead to other climatechange-related dormant commerce clause challenges. One possible object of such challenges might be California's cap-and-trade system—in particular, its requirement that importers of electricity account for their emissions. Another might be SB 1368, a 2006 California law that set an "emission performance standard" for all long-term power contracts and baseload generation. The standard was set at 1,100 pounds of CO_2 per megawatt-hour. Since most of the generation that exceeds that standard is located outside California (in the coal states of Wyoming and Montana), the law might be seen to overburden out-of-state competitors.[60]

H. The Public Trust Doctrine and GHG Emissions

In May 2011, a coordinated campaign of lawsuits and rulemaking petitions was initiated based on the argument that (1) the states and the federal government have a public trust responsibility to protect the atmosphere, and (2) with regard to climate change, they have failed to exercise that responsibility.[61] Either a lawsuit (about 12) or a petition (about 40) was filed in each state. The lawsuits and petitions, many filed by minors through their guardians ad litem, are being coordinated by Our Children's Trust, an Oregon nonprofit. As background, the public trust doctrine is an ancient common law principle with origins in Roman law and the Magna Carta. It asserts that certain natural resources are held by the sovereign in special status. Key aspects of that special status are that government may neither alienate public trust resources nor, more pertinent here, permit their injury by private parties. Rather, government has an affirmative duty to safeguard these resources for the benefit of the general public. The doctrine is generally a principle of state law, though there is limited recognition of a federal counterpart. After tidelands and the beds of navigable waterways, fish and wildlife are the natural resources most traditionally associated with the public trust doctrine; courts do not appear to have applied the doctrine to the atmosphere yet, as the suits and petitions here are seeking. As for the lawsuits, each one reportedly asks the court for declaratory relief proclaiming that the atmosphere is a public trust resource and that the government in question has a fiduciary duty as trustee to protect it. Some of the suits ask for injunctive relief as well. For example, the suit against the United States asserts that the federal government has violated its trustee duties by allowing unsafe amounts of GHGs into the atmosphere and asks for an injunction requiring it to take action "consistent with the United States government's equitable share of the global effort."[62] This suit was recently dismissed by the district court, on the grounds that (1) the public trust doctrine is a purely state law doctrine, so a federal court lacks jurisdiction, and (2) under *American Electric Power v. Connecticut*,[63] use of the public trust doctrine in the air pollution context has been displaced by the Clean Air Act.[64] In the only other known court decision on the merits so far, in Minnesota, the court held that the public trust doctrine applies only to navigable waters, not the atmosphere.[65] As for the rulemaking petitions, these have been denied in at least 27 jurisdictions.[66] The negative results of the public trust litigation and petitions thus far are not surprising. As much as because the suits and petitions seek a major expansion of the public trust doctrine, courts are traditionally reluctant to obtrude themselves into matters, such as global climate change, that go beyond the capabilities of a court to manage.

III. LIABILITY FOR HARMS CAUSED BY CLIMATE CHANGE

Based on consensus predictions as to the many harms that climate change may cause, one may safely predict that liability lawsuits will be filed. This report previously mentioned the standing hurdle looming before climate change plaintiffs, especially those that are not states, and the political question hurdle. Following are some additional issues in liability actions.

A. Liability after American Electric Power Co., Inc. v. Connecticut

In *American Electric Power Co., Inc. v. Connecticut*,[67] the Supreme Court read the CAA to bar federal judges from imposing their own limits on GHG emissions from fossil-fuel-fired power plants, separate from those imposed by EPA under that act. More formally, the Court held that the CAA displaces any federal common law of nuisance that might ground a claim seeking judicial abatement of such emissions. However, *American Electric Power* left open two key questions. First, may those suffering climate-change impacts still assert federal common law of nuisance actions seeking not injunctive relief, as plaintiffs sought in *American Electric Power*, but rather *monetary damages*? Second, do *state law* claims, either common law or statutory, withstand *American Electric Power*, which addressed only *federal* common law claims?

Recently, both these questions were answered in the negative. In *Comer v. Murphy Oil Co.*, Mississippi land owners pressed state and federal tort claims (nuisance, trespass, and negligence) against numerous oil, coal, and chemical companies that allegedly emitted substantial GHGs.[68] The land owners' claims were based on property-related harms suffered as the result of Hurricane Katrina—they argued that the defendants, through their GHG emissions and resulting climate change, had contributed to warmer ocean temperatures that had intensified the hurricane, and to rising sea level that aggravated the hurricane's impacts further. They sought damages. Despite the differences from *American Electric Power*—state rather than federal claims, monetary rather than injunctive relief—the district court had little difficulty finding that decision controlling. Here as in *American Electric Power*, the court said, the lawsuit called upon the court to determine what level of CO_2 emissions was unreasonable, a determination the Supreme Court explained had been entrusted by Congress to the EPA. Therefore, the court determined that the plaintiffs' "entire lawsuit" is displaced by the CAA,[69] though the ruling is dictum.[70]

The reach of *American Electric Power* may soon be tackled again in the appeal of the district court decision in *Native Village of Kivalina v. ExxonMobil Corp.*[71] In this case, Inupiat Eskimos forced to relocate their coastal village due to shore erosion sued 20 energy and utility companies for damages. Their claim was that the defendants' GHG emissions had, by exacerbating climate change, contributed to the melting of sea ice that had protected the village's shores from wave erosion. The district court decision, rendered prior to *American Electric Power*, rejected the claim on standing and political question grounds. These issues will be before the circuit court, of course, in addition to the displacement question under *American Electric Power*.

B. Insurance Coverage of Injury or Liability Associated with Climate Change

Federal and private insurers are well aware that if the scientific consensus is correct that climate change will bring on more frequent extreme weather events, they stand to make substantially increased payments.[72] At this time, there appear to be no insurance policies that provide explicit coverage for injuries resulting from climate change; however, there are policies that cover many of the injuries likely to be associated with climate change, "such as flood, wind, freezing, heat, earth movement, or collapse."[73]

Some issues in the vast universe of insurance-coverage litigation seem to be especially relevant to climate change. One arises from coastal hurricanes, the impacts of which may be exacerbated by climate-change-induced sea level rise. The issue is whether a particular item of hurricane damage is to be regarded as wind-caused damage or flood-caused damage. The distinction is pivotal because domestic insurance policies cover only wind damage; flood damage is insured under the National Flood Insurance Program.[74] The litigation in this area, such as that generated by Hurricane Katrina, is voluminous and often turns on factual questions, but also raises such issues as (1) who, insurer or insured, bears the burden of showing the portion of damage covered by the policy when both an insured (say, wind-caused) risk and a non-insured (say, flooding-caused) risk contributed;[75] (2) whether water driven by wind ("storm surge") falls outside the flooding exclusion in homeowners' policies;[76] and (3) whether the flooding exclusion covers man-made causes (e.g., negligent maintenance of levees) as well as natural ones.[77]

Another issue is whether the Comprehensive General Liability (CGL) policy used by businesses covers liability imposed on the insured as the result of the insured's GHG emissions, when those emissions contribute to climate-change-related damage. The only known decision on this issue is *AES Corp. v.*

Steadfast Ins. Co.,[78] a ruling by the Virginia Supreme Court that the insurance company was not obligated to provide defense under its CGL policy with AES in the *Kivalina* suit,[79] because Kivalina's complaint did not allege an "occurrence."

Finally, some policies, such as environmental liability or pollution policies, cover damage from "pollution." Where "pollution" is defined in policies to mean substances classified as pollutants under environmental laws, the Supreme Court decision in *Massachusetts v. EPA* may prove pivotal.[80] There, the Court held that GHG emissions are "air pollutants" under the Clean Air Act, raising the possibility that this ruling will be used to enlarge policy coverage to bring in damage traceable to GHG emissions.

C. U.S. Liability in International Fora Based on GHG Emissions

Whether sovereign nations may be, or should be, liable under international law for failing to reduce GHG emissions within their territory has long attracted the attention of commentators[81]— and, of course, low-lying nations. However, research fails to reveal any successful effort to impose such liability.

Some principles that might be applied to a claim alleging GHG-caused injury might be taken from the international law of transboundary pollution. For example, the Restatement (Third) of Foreign Relations Law describes an international law principle under which a nation must "take such measures as may be necessary, to the extent practicable under the circumstances, to ensure that activities within its jurisdiction or control ... are conducted so as not to cause significant injury to the environment of another state."[82] Similarly, the Trail Smelter arbitration decision, probably the seminal ruling on state liability for transboundary pollution, declared that "[a] State owes at all times a duty to protect other States against injurious acts by individuals from within its jurisdiction."[83] Of course, as with the domestic litigation, daunting hurdles confront the international-law claimant in making the link between climate change in general and specific environmental harms, and in apportioning how much of such harms to attribute to the charged parties.

Research reveals only one climate-change-related international law action filed against the United States. In 2005, the chair of the Inuit Circumpolar Conference, on behalf of herself and all affected Inuit of the arctic regions of the United States and Canada, filed a petition against the United States with the Inter-American Commission on Human Rights, the investigative arm of the Organization of American States (OAS).[84] The petition alleged that the United States, through its failure to restrict its GHG emissions and the resultant climate change, had violated the Inuit's human rights—including their rights to their culture, to property, to the preservation of health, life, and to physical integrity. Inuit culture is described in the petition as "inseparable from the condition of [its] physical surroundings."[85] Generally, the Inter-American Commission on Human Rights is empowered to recommend measures that contribute to human rights protection, request states in urgent cases to adopt specific precautionary measures to avoid serious harm to human rights, or submit cases to the Inter-American Court of Human Rights. The United States, however, has not accepted the jurisdiction of this court, so the Inuit petition sought only to have the commission prepare a report declaring the responsibilities of the United States and recommending corrective measures. In 2006, the Inuit petition was rejected, with no reasons given (as is customary for the commission).

IV. CLIMATE CHANGE-INDUCED WATER SHORTAGES

A. Water Scarcity and Water Rights

It is widely predicted that climate change will exacerbate water scarcity—widening arid areas and making them even drier. The future of the western United States has received substantial attention in this regard.[86] Where demand outstrips supply, the nature and flexibility of existing water rights are raised.

To be sure, water rights, mostly a creature of state law, are property of a uniquely conditional nature. Most obviously, the water rights holder does not own the water to which the right applies; the right is merely "usufructuary," that is, to *use* the water. In the western United States, water rights generally are governed by "prior appropriation" doctrine, under which the right of use is contingent on the right holder putting the water to "beneficial use," and is further subject to common law or statutory limits based on the public trust doctrine and the doctrine of reasonable use. With regard to "reasonable use," the California Constitution, as an example, declares that the "unreasonable use or unreasonable method of use of water be prevented," a doctrine that is self-executing and evolving.[87] Appropriation doctrine is a "first in time, first in right" system under which inadequate supply results in junior-in-time appropriators having their water cut before senior-in-time appropriators.

Despite the conditionality of water rights, it remains to be seen how much latitude government agencies have to respond to periods of water scarcity by cutting back on the consumption of vested water rights holders to accommodate critical public needs. It is also unclear to what extent appropriation doctrine states may allow water rights holders to transfer water rights, generally favored by scholars as promoting more efficient outcomes and the achieving of environmental goals.[88] One writer has noted that in the West, the explosive population growth of recent decades has often occurred in communities with only junior water rights. Senior water rights holders often include older municipalities, mining, and agriculture.[89] The question then arises whether reasonable use and other doctrines qualifying appropriation water rights can address the difficult situation of new communities being starved for water while senior appropriators endure little or no reduction in water supplies.

States have evolved a variety of additional mechanisms for allocating water among rights holders in times of scarcity. Many states exempt certain "domestic" uses of water (e.g., for stock watering, home use, or lawn watering) from the general permit scheme. If climate change produces more droughts, conflicts will increase between exempted users and those with appropriation rights, especially senior appropriators. In some cases, the ability of an exempted user to leapfrog over the rights of senior appropriators may be held subject to payment of compensation under the constitutional right to compensation for the taking of property.[90]

The issues raised above are also likely to arise in the context of groundwater, which, as with surface water, is usually held under a right of use only, not outright ownership.[91] A recent Texas Supreme Court decision adopted the minority view of outright ownership, the court reassuring that conservation of groundwater still can be done without takings as long as the problems of limited water supply "are shared by the public, not foisted onto a few."[92]

B. Water Diversion and Delivery Cutbacks

Periods of low precipitation, as may be more frequent in the future due to climate change, have generated several court decisions where the conflict was between the water needs of the public and those of fish in streams. These decisions resolved claims of Fifth Amendment takings of water rights and claims of government breach of water-supply contracts based on cutbacks in the amount of water delivered from federal water projects—as demanded by the Endangered Species Act[93] and the Central Valley Improvement Act.[94] A key issue in these cases has been whether the taking claim is to be analyzed by the court as a physical taking of the water, or as a regulatory taking of use rights in the water. The distinction matters a great deal. In general, a plaintiff's litigation prospects are substantially improved if the court adopts a physical takings framework, thus the physical versus regulatory takings issue has been hard fought in the courts. Currently, it appears that when the government requires a *physical diversion of the water* away from the plaintiff's desired use (as to operate a fish ladder), the plaintiff-friendly physical taking approach is triggered.[95] But, it would appear, not otherwise.

Another issue has been the role of doctrines that qualify water rights—principally, public trust and reasonable use.[96] Do these doctrines allow the government to set supervening public priorities for fish preservation as part of rights it retains when conferring water rights? If the government retains such rights, no taking claim can succeed, for the water rights holder cannot be found to have suffered a taking of a right he or she never acquired.

V. SEA LEVEL RISE AND EXTREME PRECIPITATION

A. Effect of Sea Level Rise on the Beachfront Owner's Property Line

Sea level rise generally causes the boundary between land and water to move landward.[97] The common law has long had to deal with such shifting boundaries—in particular, with who owns land newly dry or newly submerged. The rule, dating back to Roman times, turns on whether the land-water boundary shift occurred slowly or quickly.[98] When land-water boundaries shift *gradually and imperceptibly*—"so slowly that one could not see the change occurring"[99]—the ownership boundary shifts with it. Thus, in the case of "accretion," defined as the gradual depositing of alluvion (sand, sediment, or other deposits) so as to enlarge one's tract, the owner of the tract becomes the happy owner of the accreted area as well. The shore owner may be less pleased, however, with "erosion," the gradual and imperceptible boundary shift towards land when former upland is submerged. As with accretion, the property line moves—landward this time.[100]

In contrast with accretion and erosion, *sudden* shifts in the land-water boundary, known regardless of direction as "avulsion," do not shift ownership lines. A classic avulsive event is a hurricane that abruptly shifts the mean high water mark on a beach either seaward or landward. In this case, the property line between the owner of the intertidal zone and permanently submerged lands (typically the state in trust for the public) and the owner of uplands beyond the high water mark (typically a private entity) does not move.

The pivotal question is whether movement in the land-water boundary owing to climate-changecaused sea level rise is fast enough to be avulsive, leaving the property line unmoved, or gradual enough to be erosion, reducing the shoreowner's property.[101] No caselaw on the point exists and commentators are divided. One scholar asserts: "The rising sea level [from climate change] is neither gradual like traditional accretion, erosion, or reliction; nor is it sudden and violent like traditional avulsion. We are facing a historically distinct situation that is not a good factual fit with the [traditional common law] rules."[102] Two other scholars, in contrast, do see the requisite gradualness for property line movement: "in most instances sea level rise [from climate change] will transform private property into public property as sea waters cover formerly dry land."[103]

Case law authority does suggest that public trust ownership of coastal submerged lands and the adjacent intertidal zone (between low and high water mark) expands automatically when erosion occurs. That is, no legal process is required. In *McQueen v. South Carolina Coastal Council*, for example, that state's high court decreed that under state law, wetlands created by the encroachment of navigable tidal water belong to the state—that is, are public trust property. Proof that such lands were upland when acquired and that the tidelands were subsequently created by the rising of tidal water, said the court, cannot defeat the state's presumptive title to the tidelands.[104] As well, the court held, the state incurs no takings liability.

As long as state courts are able to ground such extensions of public trust lands in traditional common law, no Fifth Amendment taking from beachfront property owners is likely to be discerned. Title to coastal property (or any other property) is assumed to be qualified by traditional common law principles, and public trust doctrine certainly falls in this category. On the other hand, if courts use sea level rise as an occasion to expand public trust doctrine beyond its traditional state-law parameters or to otherwise shrink littoral rights, the possibility of a so-called "judicial taking" may arise. This novel concept, that *courts* may effect takings just as other branches of government do, received a major boost in 2010 when a Supreme Court plurality proposed that "[i]f a legislature *or a court* declares that what was once an established right of private property no longer exists, it has taken that property."[105] As yet, however, no court has ever found a judicial taking in a final decision.

B. "Rolling" Public Beach Access Easements

A case out of Texas is being closely watched as suggestive of constitutional issues that may be raised by landward migration of beaches from climate-change-related sea level rise. Though the case involves landward migration as the result of a hurricane, it could just as easily have arisen in connection with sea level rise (or hurricane impacts enhanced by sea level rise).

Severance v. Patterson[106] deals with the Texas Open Beaches Act, which imposes a public access easement on the state's beaches extending landward to the dune vegetation line. The lower Texas courts had long construed this access easement to "roll"—that is, to migrate with movements in the dune vegetation line. The consequence is that landward movement of the vegetation line may result in private land, including improved parcels, being newly encumbered by the easement. Under the act, the state may then order the house removed, although some compensation is provided for removal expenses. Carol Severance bought two houses behind the vegetation line, only to have Hurricane Ike a few months later move the line landward of her houses—making them subject to removal orders. She asserted Fifth Amendment takings and Fourth Amendment unreasonable seizure claims.

The Fifth Circuit found the taking claim unripe, but certified questions to the Texas Supreme Court as to Severance's Fourth Amendment claim. In its answers, the Texas Supreme Court narrowed the circumstances when the public access easement rolls.[107] It concluded that "[a]lthough existing public easements in the dry beach of Galveston's West Beach are dynamic, as natural forces cause the vegetation and the mean high tide lines to move gradually and imperceptibly, these easements do not spring or roll landward ... as a result of avulsive events." In so ruling, the court reversed the decades-old interpretation of the Texas Open Beaches Act in the lower state courts, which had allowed the public access easement to roll no matter how abrupt the movement in the vegetation line. Also important, the Texas court ruling raises again the question asked in Section IV.A. as to whether climate-change-caused sea level rise should be considered gradual or avulsive.[108]

Another often-cited example of statutes anticipating landward migration of beaches are the coastal sand dune rules promulgated by a Maine state agency under that state's Natural Resources Protection Act.[109] The rules bar a project in a coastal sand dune system "if, within 100 years, the project may ... be eroded as a result of changes in the shoreline such that the project is likely to be severely damaged after allowing for a two foot rise in sea level over 100 years."[110]

C. Shifting Floodplain Designations

Sea level rise and extreme rains born of climate change may cause lands not formerly subject to flooding to become so. Land use planners have long encountered resistance updating floodplain designations because such a designation alerts potential buyers that a parcel is vulnerable, possibly reducing the parcel's market value. It is unlikely, however, that a floodplain designation could, in itself, result in enough value loss to constitute a Fifth Amendment regulatory taking of a property.[111]

D. Levee-Related Issues

Damage from climate-change-caused extreme weather or sea level rise may require courts in the future to clarify federal liabilities in connection with Army Corps of Engineers levee construction and operation. The extensive litigation following the breaching and overtopping of the levees protecting New Orleans during Hurricane Katrina may be a harbinger of climate-change-related litigation in the future. CRS Report RL34131 examines the potential liability of the United States in connection with Hurricane Katrina and other flooding—discussing the Federal Tort Claims Act, the Flood Control Act of 1928, and negligence theory.[112]

A quite recent decision holds that the Corps of Engineers' negligent maintenance of a shipping channel between New Orleans and the Gulf of Mexico had the effect of channeling Hurricane Katrina storm surge to the city, breaching levees. Accordingly, the court imposed tort liability on the United States.[113] Contrariwise, a takings claim based on Katrina-related damage to New Orleans, alleging the Corps' failure to adequately design, build, or maintain adequately the levees themselves, was rejected.[114] The gist of these and other decisions is that while the government has no duty to protect the public and its property from flooding, liability may be imposed where government structures worsen floods.[115]

Separate issues have been raised by the Corps' *intentional* releases of floodwaters from the Mississippi River in May 2011, following unusually heavy rainfalls combined with raised water levels due to snowmelt. These issues, mostly concerning the Corps' authority to release waters intentionally and the adequacy of the flowage easements obtained by the Corps, are also treated in CRS Report RL34131. The adequacy of the flowage easements obtained from landowners by the Corps, in advance of the intentional releases, is front and center in two pending class action complaints arising from the releases, claiming takings.[116]

Beyond flowage easement issues, the intentional-flooding litigation poses the question whether the flooding should be analyzed as a potential tort or instead as a potential Fifth Amendment taking of property rights. Here there is a climate-change-related twist. Under long-established case law, the distinction between a flood that is a tort and one that is a taking turns on whether the flooding, if not permanent, is at least "inevitably recurring."[117] If inevitably recurring, the flood is to be analyzed as a possible taking, specifically as a taking by permanent physical occupation, and jurisdiction vests in the Court of Federal Claims. If not, the flood is at most a tort and jurisdiction lies in the district court.[118] In both the Mississippi River class actions, the United States has filed motions to dismiss alleging that the rarity of intentional releases like those in the case demands that they be regarded as not "inevitably recurring," hence at best a tort. But the scientific consensus asserts that as climate change progresses, extreme precipitation events such as those at issue here may become more common. If that happens, will the United States be able to assert that future intentional releases of floodwaters following heavy precipitation are not inevitably recurring, hence are at most a tort? The distinction is of some moment, given that legal defenses available to the United States in the event of a tort are not available in the event of a taking.[119]

A final levee-related issue is suggested by a recent news article describing opposition of residents in Virginia's Middle Peninsula to planners' proposal to rezone land for use as a dike against rising water, and noting that "[o]utside of greater New Orleans, Hampton Roads is at the biggest risk from sea-level rise of any area its size in the United States."[120] Again, the spectre of takings claims looms if the rezoning results in the severe devaluation of parcels, or is analyzed as a physical taking.

E. Failure to Take Preventive Measures

The scientific consensus that climate change will lead to further sea level rise raises the issue whether governments can be held liable for failing to act to avert the harmful impacts of such rise. Generally, failure to act cannot be the basis of a taking claim. But when a city fails to act on a hazard that is specific and well understood, negligence may lie. Thus, in one case with relevance to future heavy rains from climate change, the court held that allegations that a city was aware of the potential for overflow from the city landfill's retention ponds, and its subsequent failure to take measures to prevent such overflow, did not state a taking claim, but did properly assert negligence.[121]

VI. OTHER ADAPTATION RESPONSES TO CLIMATE CHANGE

The previous section touched on a few adaptation measures specifically related to sea level rise. This section continues with additional adaption measures that raise legal issues.[122]

A. Beach Issues

1. Armoring

Shoreline "armoring"—seawalls, revetments, and bulkheads[123]—has obvious relevance to climate-change-caused sea level rise. The definition of armoring in the Florida administrative code is as good as any: "a manmade structure designed to either prevent erosion of the upland property or protect eligible structures from the effects of coastal wave and current action."[124] The right to erect shore defense structures on one's property has long-standing common law imprimatur, yet the practice has its detractors. Seawalls, for example, have been said to deflect waves onto other beaches, causing sand to be scoured away, and also to cut off the natural supply of sand to the beach from the sand dune behind the wall.

Many have proposed that states adopt anti-armoring statutes, so as to allow the natural landward migration of the land-water boundary caused by sea level rise. Such natural migration of the boundary allows the creation of new, ecologically valuable wetlands to replace those lost to sea level rise, and the expansion of public trust lands. An obvious issue, however, is whether these consequences of anti-armoring laws trench on private property rights in a manner that must be compensated as a taking. Though the issue is certainly unresolved by the limited relevant litigation, the balance of arguments seems to tip against a taking. Most obviously, the harm to the littoral owner (from flooding and encroaching public trust lands) likely would be viewed by courts as resulting from sea level rise, not the armoring restriction.[125]

A taking claim was rejected, logically enough, where the shore owner proposed armoring on public trust lands. The case is *McQueen v. South Carolina Coastal Council*,[126] in which the state denied the owner of a tract along a manmade canal permission to build a seawall and to backfill. Even though without the seawall the tract was assumed to be unbuildable and have zero value, no taking of plaintiff's property was found to have occurred. As the court saw it, plaintiff's land had largely reverted to public-trust tideland belonging to the state by the time his application was denied. Thus, the seawall permission denial took nothing plaintiff had at the time of his application. Recall the earlier discussion of shifting public trust in connection with this case in section IV.A.

In the absence of armoring restrictions, one can expect sea level rise to cause more beachfront land owners to install defensive structures. As a result, questions as to liability for harm to neighboring tracts may be raised more often. A hoary common law principle, the "common enemy doctrine," holds that one may erect defenses against the sea even though doing so may cause water to beat with added force against adjoining lands and require the adjoining landowner to also erect defenses.[127] Many states, however, have moved away from the common enemy doctrine toward a rule of reasonableness, under which liability for harm to others is avoided only when the interference with the flow of surface waters is "reasonable," a term that could benefit from judicial clarification.[128]

One case takes on the tantalizing question of whether armoring structures block the landward shift of the line between public and private ownership, typically the mean high water mark, when that mark reaches such a structure. In *United States (Lummi Nation) v. Milner*,[129] the Ninth Circuit said no; the ownership line continues to move as if the armoring structure had not been built. While the upland owner has the right to erect structures on his or her property to defend against erosion and storm damage, the tideland owner has "a vested right to the ambulatory boundary and to the tidelands they would gain if the boundary were allowed to ambulate."[130] In short, the upland owner "[does] not have the right to permanently fix the property boundary" absent the tideland owner's consent.[131] The court pointed out that its ruling might have limited applicability, given that the tideland owner here was an Indian tribe and its federal trustee, rather than the state as in the usual case. This allowed the federal court to create federal common law, while most such disputes over tideland/upland boundaries are handled by state courts under state law. One commentator notes that "[t]he decision, if applied generally, might make many homes now behind seawalls trespassers on state property."[132]

2. Renourishment

Adding sand back to eroded beaches or building up beaches, often called beach "nourishment" or "renourishment," may be increasingly resorted to as climate change progresses and sea level rises. In the near term (but unlikely beyond), repairing the ravages of storms may be preferable to the difficulties of moving existing coastal population inland. Even Members of Congress who generally seek to limit federal spending have strongly supported Corps of Engineers beach restoration projects where the local economy depends on attractive beaches.[133]

The Supreme Court, too, has turned its attention recently to beach renourishment projects. In *Stop the Beach Renourishment, Inc. v. Florida Dep't of Environmental Protection*,[134] the Court confronted a Florida beach renourishment project that had sparked objections from a handful of the affected beachfront property owners. Those owners insisted that by adding a strip of state-owned beach in front of their eroded privately owned beach, the state had effected a Fifth Amendment taking of two of their littoral property rights: the right to ownership of future accreted land and the right to direct contact with the water. The Supreme Court held unanimously that the Florida Supreme Court had properly found no taking, since the shore owners had not shown that these littoral rights were superior to the state's right to fill in its submerged land. Note that the restored beach belonged to the state: "Florida law as it stood before the decision below allowed the state to fill in its own seabed, and the resulting sudden exposure of previously submerged land was treated like an avulsion for purposes of ownership."[135] Avulsions, recall, do not move ownership boundaries.

While *Stop the Beach Renourishment* was a victory for beach renourishment efforts, the decision turned on Florida case law precedent that may not be replicated in other states. Thus, legal challenges by littoral owners to beach restoration projects can be expected to continue.

Because the Florida and U.S. supreme courts found no property rights impaired in *Stop the Beach Renourishment*, they had no occasion to clarify how the *benefit* to the beachfront property owner from renourishment might factor into the taking analysis. This is a pivotal question if the costs of beach renourishment are to remain affordable, but two recent state court decisions give opposite answers. In one, a New Jersey court confronted a municipality's condemnation of an easement to erect a 22-foot-high dune on beachfront property, to protect the barrier island from storms.[136] The court held that the jury's $375,000 compensation award, largely for the dune's partial blockage of the ocean view, was not to be reduced by the storm-protection benefit conferred on the property owner. Under well-established law, the court said, compensation awarded a condemnee is offset only by benefits of the project specific to the condemnee ("special benefits"), not those enjoyed by the community at large ("general benefits"). The benefit conferred by the dune was protection of the island from storms—in the court's view, a general benefit, hence not an offset. A contrary view comes from a North Carolina court in a case where a state agency offered zero compensation for an easement over private beachfront property needed to implement a beach renourishment project. The court found the offer reasonable (though subject to final determination in an eminent domain proceeding), citing the renourishment project's benefits to the beachfront property owner as adequate compensation.[137]

B. "Retreat"—Moving Development Inland

Levees, armoring, and beach restoration, discussed above, have long been well-understood techniques, widely supported by land owners if not by environmentalists. Given sea level rise of the magnitude predicted in connection with climate change, however, the long-term viability of such structural protections seems dubious.[138] Attention is shifting instead toward "retreat"—an unfortunately pejorative term denoting government actions that discourage new development in disaster-prone areas (proactive retreat) or reconstruction following such disasters (reactive retreat).[139] When that discouragement takes the form of outright regulatory prohibition—rather than merely removal of development incentives—the taking issue is likely to arise yet again.[140]

The legal question is whether the specific context of sea level rise due to climate change may offer the government defenses against regulatory takings claims not otherwise available. One possible starting point is *Lucas v. South Carolina Coastal Council*.[141] There, the Supreme Court dealt with a state beachfront management act aimed in large part at protecting the beach/dune system along the state's coast. Toward that end, the act sought to "discourage[e] new construction in close proximity to the beach/dune system and encourag[e] those who have erected structures too close to the system to retreat from it."[142] In particular, the plaintiff was barred from building any occupiable structure on his two beachfront lots. The Court pointedly rejected the state's assertion that the statute, by asserting avoidance of a public harm as its purpose, was immunized from takings liability. Only state action based on "background principles of the State's law of property and nuisance" was so protected,[143] said the Court, holding that the beachfront management act did not fall into that category. Traditional common law, it observed, rarely supports prohibiting the erection of a house.[144]

Lucas suggests that the possibility that a tract of land will be submerged in the future as the result of climate change may not be sufficient to deflect takings or other legal challenges against a development prohibition on that tract—at least when, as in *Lucas*, the prohibition eliminates all land value. In *Lucas*, not even the fact that plaintiff's lots had been submerged at various times in the previous 40 years was enough to shield the state from takings liability. And while public trust doctrine has been held to be a "background principle" immunizing the state,[145] there is no support for any extension of public trust doctrine, as a defense to takings claims, to lands not below the mean high water mark when the development prohibition is imposed. Arguably, however, the question remains open.[146]

The *Lucas* decision, rendered in 1992, did not consider climate change. And because *Lucas* dealt with a "total taking"—that is, a regulatory restriction eliminating *all* use and value in a tract of land—it did not deal with takings law factors confined to *less-than-total* elimination of use and value. One such factor is the extent to which the government action interfered with the landowner's "reasonable investment-backed expectations" (RIBEs).

The RIBEs question here revolves around recent or future purchasers of land prone to climatechange-induced extreme weather, such as flooding. Can such purchasers be charged with constructive knowledge of the scientific consensus that climate change will bring about more frequent instances of extreme weather in the future? Can such purchasers, as a result, be held "on notice" that state or local governments might restrict development of such parcels in the future, weakening any claim that such restrictions interfere with *reasonable* expectations of development when the land was acquired? Would the existence of a widely publicized government retreat proposal at the time when the land was acquired strengthen an on-notice/absence-of-RIBEs argument by the government? And could states bolster this defense by requiring that all purchasers of disaster-prone land be given written notice prior to purchase of the risks to which they were exposing themselves?[147] Even today, "[s]everal [state] disclosure statutes require inclusion of whether the property has been affected by floods or is in a flood zone or plain."[148] The extremely thin case law on whether such notice undercuts a taking claim based on development restrictions points to notice not making much difference.[149] But it is far too early to regard the matter as settled.

The question has also been raised whether local jurisdictions might be successfully sued in the opposite situation—that is, where they *fail* to restrict development despite having knowledge that flooding may occur, following which the permitted development is damaged by flooding or exacerbates flooding on other properties.[150]

Further inland, the National Flood Insurance Program (NFIP) becomes a central player in discouraging construction in flood-prone areas[151]—floods that in some instances may become more severe or frequent as the result of climate change-related sea level rise or extreme rainfall.

A local jurisdiction's participation in the NFIP is voluntary. It is embodied in an agreement under which the community adopts floodplain management ordinances meeting minimum NFIP requirements for regulating new-construction design in "special flood hazard areas,"[152] and use restrictions in the regulatory floodway. In return, the federal government makes subsidized federal flood insurance available to landowners in those jurisdictions.

Courts have rejected unanimously takings suits based on NFIP-inspired floodplain ordinances, or similar non-NFIP floodplain ordinances.[153] But the NFIP does not now account for future sea level rise. Should future sea level rise lead to stricter federal conditions for flood insurance in the form of stricter floodplain ordinances, takings issues inevitably will rear their head once more. One can expect, however, that the current judicial refusal to impute to the United States any takings liability for such local ordinances will continue to stand as long as their adoption remains voluntary.[154]

Finally, local jurisdictions have asked whether their potential disinvestment in public infrastructure in low-lying areas (such as armoring, roads, and wastewater treatment plants) might raise takings issues.[155] The aim of such disinvestment would be to hold down flood-induced costs by discouraging new development in such areas or stimulating removal of existing development.

Affected property owners, however, may not be so civic-minded. For example, a state's decision to discontinue maintenance of a shoreside road that is eroding away might lead those dependent on that road for access to their property to assert a taking by denial of access.[156] The viability of such takings claims will vary widely with the facts.

No reported takings decisions at all exist in response to the federal government's disinvestment in the development of coastal barrier islands through the Coastal Barrier Resources Act.[157] On the other hand, disinvestment in public infrastructure may be dicier, if the courts perceive a state or local government duty to maintain existing infrastructure.[158] Presumably, takings problems can be lessened by announcing disinvestment many years (even a decade or more) in advance; such "amortization periods" have been effective in other factual contexts, such as billboard removal programs, in deflecting takings claims.[159] Governments might also take care not to allow disinvestment in an area to get too far ahead of the retreat activity of those living and working there.

VII. RESPONDING TO AND REBUILDING AFTER NATURAL DISASTERS

A. Responding

Legal questions inevitably arise as to whether public and private actions taken in an emergency, climate-change-related or otherwise, are subject to the same legal requirements as when there is no emergency. And, for that matter, what constitutes an emergency—a term generally left undefined in statutes. There is no explicit, across-the-board exemption in any federal environmental law for emergency response.

A sampler of less-than-across-the-board provisions reflecting the need for expedition in emergencies might include, first, the Superfund Act.[160] Under this act, government response to releases or threatened releases of hazardous substances, as when a flood jeopardizes containment of hazardous chemicals at a site, can be done hurriedly as emergency actions (known as "removal actions") with less prior study and investigation than is required for permanent cleanups (known as "remedial actions").

Similarly, Council on Environmental Quality regulations implementing the National Environmental Policy Act (NEPA) say that where emergency circumstances require a federal agency to take action without observing the regulations, the agency should consult with the Council about "alternative arrangements."[161] Federal actions not needed to control the immediate impacts of the emergency, however, remain fully subject to NEPA review.

B. Rebuilding

Following a natural calamity in which structures are destroyed, questions often arise whether the rebuilding of a structure essentially as it was before, in the very same location, is subject to the full range of environmental and other requirements applicable if the structure were being built there for the first time. Here, besides the question of what constitutes an emergency, there is the added issue whether the replacement structure is essentially the same as its predecessor (changes are always made to some degree). As with responding to emergencies (previous section), there appears to be no explicit, across-the-board exemption in federal environmental law.

Probably the broadest exemption in federal statutes for rebuilding structures is that in the Stafford Disaster Relief Act. The act decrees that no environmental impact statement (EIS) under NEPA is required for "[a]n action which is taken or assistance which is provided pursuant to [the Act], which has the effect of restoring a facility substantially to its condition prior to the disaster or emergency."[162] Also as to NEPA, Department of Transportation regulations allow for categorical exclusions from EIS preparation for reconstruction (whether prompted by a disaster or not) of highways, bridges, and rail and bus facilities.[163]

Limited NEPA case law on the replacement issue indicates that federal involvement in the construction of an essentially similar replacement facility does not require an EIS—as long as the environment *with* the original facility is accepted as the status quo baseline.[164] This qualifier suggests that the passage of several years before the new facility is built, accompanied by a change in the environment at the site, might cause the *changed* environment to be viewed as the baseline. With the changed environment as the baseline, the federal action might be seen as having significant impact, triggering the EIS requirement.

Outside of NEPA, the Clean Water Act affords an exemption from its requirement of permits for the discharge of dredged or fill material "for the purpose of emergency reconstruction ... of currently serviceable structures such as dikes, dams, levees, ... and transportation structures."[165] Also, three nationwide permits issued by the Corps of Engineers under this permit program cover reconstruction in varying degrees, relieving the applicant of the more expensive and time-consuming process of applying for an individual permit.[166]

VIII. IMMIGRATION AND REFUGEE LAW[167]

United Nations High Commissioner for Refugees Antonio Guterres has said: "Climate change is today one of the main drivers of forced displacement, both directly through impact on environment—not allowing people to live any more in the areas where they were traditionally living—and as a trigger of extreme poverty and conflict."[168] Climate-related migrants, however, are not considered a "protected class" of people in international law or U.S. immigration law, nor is there a specific legal framework or entity responsible for their displacement.

In international law, the foundational document is the 1951 Convention Relating to the Status of Refugees, which defines "refugee" as a person with a "well-founded fear of being persecuted for reasons of race, religion, nationality, membership of a particular social group, or political opinion"[169]—unlikely, it would seem, to embrace climate change refugees. Similarly, the United States has long held to the principle that it will not return a foreign national to a country where his life or freedom would be threatened, but this principle does not encompass economic or environmental migrants. The Immigration and Nationality Act (INA) requires foreign nationals seeking asylum or refugee status to demonstrate a well-founded fear that, if returned home, they will be persecuted based upon the five characteristics listed in the Convention (above).[170] Provisions also exist in the INA to offer temporary protected status (TPS) or relief from removal when natural disasters occur or when violence and civil unrest erupt in spots around the world. While TPS may benefit people stranded in the United States because of natural disasters, it is only short-term relief from removal.[171]

End Notes

[1] *See, e.g.,* Michael B. Gerrard (ed.), GLOBAL CLIMATE CHANGE AND U.S. LAW (American Bar Ass'n 2007); Tom Mounteer (ed.), CLIMATE CHANGE DESKBOOK

(Envtl. Law Inst. 2009); CRS Report R40556, *Market-Based Greenhouse Gas Control: Selected Proposals in the 111th Congress*, by Brent Yacobucci and Jonathan Ramseur.

[2] National Research Council, ADVANCING THE SCIENCE OF CLIMATE CHANGE 2 (2010). See generally CRS Report RL33849, *Climate Change: Science and Policy Implications*, by Jane A. Leggett.

[3] National Research Council, *supra* note 2, at 2.

[4] *See, e.g.*, Lujan v. Defenders of Wildlife, 504 U.S. 555, 560 (1992).

[5] *See, e.g.*, Native Village of Kivalina v. ExxonMobil Corp., 663 F. Supp. 2d 863, 880 (N.D. Cal. 2009) ("there is no realistic possibility of tracing any particular alleged effect of global warming to any particular emissions by any specific person ..."), *appeal pending before Ninth Circuit*.

[6] 549 U.S. 497 (2007).

[7] 582 F.3d 309 (2d Cir. 2009), *reversed on other grounds*, 131 S. Ct. 2527 (2011) (affirming the Second Circuit's finding of standing by equally divided vote).

[8] *See generally* Kirsten Engle, *State Standing in Climate Change Lawsuits*, 26 J. Land Use & Envtl. L. 217 (2011).

[9] 549 U.S. at 520.

[10] *Massachusetts*, 549 U.S. at 518-520; Connecticut v. American Elec. Power Co., 582 F.3d 309, 338-339 (2d Cir. 2009), *reversed on other grounds*, 131 S. Ct. 2527 (2011). That is, Article III is satisfied when a state brings suit as *parens patriae* on behalf of its citizens. *Massachusetts*, 549 U.S. at 519-521. *Parens patriae* doctrine allows a state to sue in its sovereign capacity to protect its citizenry, rather than being limited, as Article III would normally require, to asserting traditional particularized injuries to state interests. The modern origins of the doctrine lie in two century-old nuisance cases brought by states in federal court alleging interstate pollution: *Missouri v. Illinois*, 180 U.S. 208 (1901), and *Georgia v. Tennessee Copper Co.*, 206 U.S. 230 (1907). In both cases, state standing was found. The current test for *parens patriae* standing is found in *Snapp & Son, Inc. v. Puerto Rico*, 458 U.S. 592 (1982), though there is some question whether traditional Article III standing requirements have to be met as well by the citizens of the state asserting *parens patriae* standing.

[11] *See, e.g.*, Comer v. Murphy Oil USA, Inc., 2012 Westlaw 933670 (S.D. Miss. Mar. 20, 2012) (finding of Article III standing for state sovereign in *Massachusetts v. EPA* does not support standing for private plaintiffs here); *Native Village of Kivalina*, 663 F. Supp. 2d at 882 (same).

[12] *See generally* Elizabeth Ann Kronk, *Effective Access to Justice: Applying the Parens Patriae Standing Doctrine to Climate Change-Related Claims Brought by Native Nations*, 32 Pub. Land & Res. L. Rev. 1 (2011).

[13] 663 F. Supp. 2d 863 (N.D. Cal. 2009).

[14] 369 U.S. 186, 216 (1962).

[15] Two decisions rejecting common-law claims based on climate-change harms, on political question grounds, are *Native Village of Kivalina*, 63 F. Supp. 2d at 871-877, and *Comer*, 2012 Westlaw 933670 *11-14. Both decisions based their rejection of the claims on the second and third *Baker* factors noted in the text. Declining to accept a political question defense for such claims is *American Electric Power v. Connecticut*, 582 F.3d 309, 323-332 (2d Cir. 2009), *reversed on other grounds*, 131 S. Ct. 2527 (2011). In contrast, no difference of judicial opinion exists when a climate change claim is based on EPA's failure to satisfy requirements in a *statute*, such as the Clean Air Act. There, the claim avoids the absence of clear standards in the common law cases and dismissal on political question grounds is

deemed inappropriate. *See, e.g., Massachusetts,* 549 U.S. at 516 (proper construction of a congressional statute, here the Clean Air Act, is a question "eminently suitable to resolution in a federal court").

[16] 549 U.S. 497 (2007).
[17] CAA §165(a)(4); 42 U.S.C. §7475(a)(4).
[18] *Id.*
[19] 42 U.S.C. §7521(a)(1).
[20] Coalition for Responsible Regulation, Inc. v. EPA, 2012 Westlaw 2381955 (D.C. Cir. June 26, 2012).
[21] 42 U.S.C. §7411.
[22] *See* 77 Fed. Reg. 22392 (April 13, 2012) (proposed standards).
[23] CRS Report R40984, *Legal Consequences of EPA's Endangerment Finding for New Motor Vehicle Greenhouse Gas,* by Robert Meltz.
[24] 42 U.S.C. §7408.
[25] 42 U.S.C. §7415.
[26] This section of the report was written by Kristina Alexander, Legislative Attorney, CRS American Law Division.
[27] Habitat change can constitute a "take" of listed species as follows. Under the ESA, "take" is defined as "to harass, harm, pursue, hunt, shoot, wound, kill, trap, capture, or collect, or to attempt to engage in any such conduct." 16 U.S.C. §1532(19). "Harm" in this definition has been defined by the Fish & Wildlife Service to include "significant habitat modification or degradation where it actually kills or injures wildlife." 50 C.F.R. §17.3.
[28] *See, e.g.,* Greater Yellowstone Coalition, Inc. v. Servheen, 665 F.3d 1015 (9th Cir. 2011); In re Polar Bear Endangered Species Act Listing, 794 F. Supp. 2d 65 (D.D.C. 2011); Center for Biological Diversity v. Lubchenco, 758 F. Supp. 2d 945 (N.D. Cal. 2010).
[29] *See, e.g.,* Conservancy of Southwest Florida v. U.S. Fish and Wildlife Service, 2011 WL 1326805 (M.D. Fla. April 06, 2011); Alliance for Wild Rockies v. Lyder, 728 F. Supp. 2d 1126 (D. Mont. 2010).
[30] *See, e.g.,* Center for Biological Diversity v. Salazar, 804 F. Supp. 2d 987 (D. Ariz. 2011); South Yuba River Citizens League v. National Marine Fisheries Service, 725 F. Supp. 2d 1247 (E.D. Cal. 2010); and Pacific Coast Federation of Fishermen's Associations v. Gutierrez, 606 F. Supp. 2d 1122 (E.D. Cal. 2008).
[31] See ESA §4(a)(1)(A), 16 U.S.C. §1533(a)(1)(A) (when making a determination on whether to list a species, the relevant wildlife agency must consider "the present or threatened destruction, modification, or curtailment of its habitat or range"; ESA §4(b)(2), 16 U.S.C. §1533(b)(2) (requiring the relevant wildlife agency to designate critical habitat); and ESA §7(a)(2), 16 U.S.C. §1536(a)(2) (requiring all agencies to consult with the relevant wildlife agency to determine whether their actions would "result in the destruction or adverse modification of habitat of such species which is determined ... to be critical").
[32] Palm Beach County Environmental Coalition v. Florida, 651 F. Supp. 2d 1328 (S.D. Fla. 2009). Plaintiffs also had alleged violations of the Clean Air Act, National Environmental Policy Act, and the Clean Water Act.
[33] 73 Fed. Reg. 76272 (Dec. 16, 2008) (effective Jan. 15, 2009).
[34] 74 Fed. Reg. 20421 (May 8, 2009) ("With this final rule, the Department of the Interior and the Department of Commerce amend regulations governing interagency cooperation under [the ESA]. In accordance with the statutory authority set forth in the 2009 Omnibus Appropriations Act (P.L. 111-8), this rule implements the regulations that were in effect immediately before the effective date of the regulation issued on December 16, 2008").

[35] 73 Fed. Reg. at 47872.

[36] 50 C.F.R. §402.02.

[37] 50 C.F.R. §17.40(q)(4). The polar bear was listed under the act primarily due to shrinking habitat caused by changing climate. 73 Fed. Reg. 28,212 (2008). The polar bear regulation prevents a lawsuit claiming that a power plant in any state other than Alaska harmed the polar bear by indirectly causing its ice floe habitat to diminish. The law that authorized revocation of the regulations discussed above, P.L. 111-8, also authorized revocation of the polar bear rule, but the Secretary of the Interior and the Secretary of Commerce did not act on that authority to revoke the rule.

[38] U.S. Const. amend. V: "[N]or shall private property be taken for public use, without just compensation."

[39] 33 C.F.R. §320.4 ("The decision whether to issue a permit will be based on an evaluation of the probable impacts, including cumulative impacts, of the proposed activity and its intended use on the public interest.").

[40] 42 U.S.C. §§4321-4370f.

[41] *See* NEPA §102(2)(C); 42 U.S.C. §4332(2)(C).

[42] City of Los Angeles v. National Highway Traffic Safety Admin., 912 F.2d 478 (D.C. Cir. 1990).

[43] *See, e.g.,* Center for Biological Diversity v. National Highway Traffic Safety Admin., 508 F.3d 508, 550 (9th Cir. 2007) ("The impact of greenhouse gas emissions is precisely the kind of cumulative impacts analysis that NEPA requires agencies to conduct.").

[44] CEQ, Memorandum for Heads of Federal Departments and Agencies, *Draft NEPA Guidance on Consideration of the Effects of Climate Change and Greenhouse Gas Emissions* (Feb. 18, 2010). As to a proposed project's possible contribution to climate change, the guidance states that "where a proposed Federal action that is analyzed in an [environmental assessment] or EIS would be anticipated to emit GHGs to the atmosphere in quantities that the agency finds may be meaningful, it is appropriate for the agency to quantify and disclose its estimate of the expected annual direct and indirect GHG emissions in the environmental documentation for the proposed action." *Id.* at 2. As to a proposed project's potential for being affected by future climate change, the guidance is equally unequivocal: "CEQ proposes that agencies should determine which climate change impacts warrant consideration in their [environmental assessments] and EISs because of their impact on the analysis of the environmental effects of a proposed agency action." *Id.* at 6.

[45] NEPA §102(2)(C); 42 U.S.C. §4332(2)(C).

[46] Compare *Sunnyvale West Neighborhood Ass'n v. City of Sunnyvale City Council*, 190 Cal. App. 4th 1351, 1382-1383 (2010) and *Madera Oversight Coalition, Inc. v. City of Madera*, 199 Cal. App. 4th 48, 90 (2011), each finding use of future conditions as the baseline to be improper, with *Neighbors for Smart Rail v. Exposition Metro Line Construction Authority*, No. B232655 (Cal. App. filed April 17, 2012), holding that in a proper case, use of projected conditions may be appropriate way to measure impacts project will have on traffic, air quality, and GHG emissions.

[47] This section of the report was written by Adam Vann, Legislative Attorney, CRS American Law Division.

[48] For a detailed discussion of these issues, see CRS Report 34307, *Legal Issues Associated with the Development of Carbon Dioxide Sequestration Technology*, by Adam Vann and Paul W. Parfomak.

[49] CAA §209(a); 42 U.S.C. §7543(a).

[50] CAA §209(b); 42 U.S.C. §7543(b).

[51] CAA §177; 42 U.S.C. §7507.
[52] 49 U.S.C. §32902(a).
[53] 49 U.S.C. §32919.
[54] Metropolitan Taxicab Bd. of Trade v. City of New York, 633 F. Supp. 2d 83, *aff'd as to EPCA*, 615 F.3d 152 (2d Cir. 2010), *cert. denied*, 131 S. Ct. 1569 (2011).
[55] As the Supreme Court noted in *Massachusetts v. EPA*, 549 U.S. 497, 519 (2007): "Massachusetts ... cannot negotiate [a GHG] emissions treaty with China or India ...." The leading decision on foreign policy preemption is *American Insurance Ass'n v. Garamendi*, 539 U.S. 396 (2003).
[56] *See, e.g.,* Green Mountain Chrysler Plymouth Dodge v. Crombie, 508 F. Supp. 2d 295 (D. Vt. 2007) (no foreign policy preemption found of Vermont's GHG emission standards for new automobiles).
[57] 2011 Westlaw 6934797 (E.D. Cal. Dec. 29, 2011), *appealing pending before Ninth Circuit*. By separate order, the court concluded that CAA §211(c)(4)(B) failed to give California immunity from dormant commerce clause challenge. The court declined for the moment to decide plaintiffs' preemption argument.
[58] U.S. Const. art. I, §8, cl. 3.
[59] Global Warming Solutions Act of 2006 (AB 32), Cal. Health & Safety Code §38,500 et seq. The LCFS regulation is at Cal. Code Regs. tit. 17, §§95,480-95,490.
[60] *See* Debra Kahn, *Traders worry that a Calif. low-carbon fuels decision could apply to electricity imports*, E&E Climatewire (Jan. 20, 2012).
[61] For further details, see *Common Law Climate-Change Litigation After* American Elec. Power Co. v. Connecticut, by Robert Meltz (CRS Report R41496).
[62] Alec L. v. Jackson, No. 11-cv-2203 (N.D. Cal. filed May 4, 2011) (complaint at 17). In December, 2011, the case was transferred to the U.S. District Court for the District of Columbia.
[63] 131 S. Ct. 2527 (2011). *See* Section III.A. of this report.
[64] Alec L. v. Jackson, No. 11-cv-02235 (D.D.C. May 31, 2012).
[65] Aronow v. Minnesota Dep't of Pollution Control, No. 62-CV-11-3952 (Minn. D. Ct. Jan. 30, 2012).
[66] *See* http://climatelawyers.com/post/2012/02/04/Aronow-v-Minnesota-is-Dismissed-Public-Trust-Doctrine-NotExtended-to-the-Atmosphere-in-Minnesota.aspx.
[67] 131 S. Ct. 2527 (2011).
[68] 839 F. Supp. 2d 849 (S.D. Miss. 2012).
[69] *Id.* at *14.
[70] When a court rules on an issue not required for resolution of the case, the ruling is referred to as "dictum." Traditionally, dictum is entitled to less precedential force than a pronouncement of the court essential to disposing of the case—often termed a "holding." In *Comer*, the *American Electric Power* discussion described in text above was preceded by not one, but three, different determinations of the court (res judicata, absence of standing, and nonjusticiale political question) each one of which was fully adequate to support dismissal of the action. That is, the court had no need to resolve the displacement issue and its discussion is, therefore, dictum.
[71] 663 F. Supp. 2d 863 (N.D. Cal. 2009), *appeal pending before Ninth Circuit*.
[72] Evan Lehmann, *Reinsurers press Congress to reduce U.S. risk from climate change (E&E ClimateWire* Mar. 2, 2012); Evan Lehmann, *Disasters, continuing to climb, inflict record insurance losses in 2011* (E&E ClimateWire Jan. 5, 2012); Government Accountability Office, *Climate Change: Financial Risks to Federal and Private Insurers in Coming*

Decades Are Potentially Significant, GAO-07-760T (2007). For general background, see Gary S. Guzy, "Insurance and Climate Change," in Michael B. Gerrard (ed.), GLOBAL CLIMATE CHANGE AND U.S. LAW (ABA 2007); Justin Pidot, Georgetown Envtl. Law and Policy Inst., *Coastal Disaster Insurance in the Era of Global Warming* (2007) (copy on file with author); Adam Riedel, *California, New York and Washington to Require Insurers to Provide Information on Climate Change Risks*, available at blogs.law.columbia.edu/climate change/2012/02/06.

[73] *Guzy, supra* note 72, at 554.
[74] 42 U.S.C. §§4001-4029.
[75] *See, e.g.*, Bayle v. Allstate Ins. Co., 615 F.3d 350 (5th Cir. 2010).
[76] *See, e.g.*, Leonard v. Nationwide Ins. Co., 499 F.3d 419 (5th Cir. 2007).
[77] *See, e.g.*, In re Katrina Canal Breaches Litigation, 495 F.3d 191 (5th Cir. 2007).
[78] 715 S.E.2d 28 (Va. 2011), *motion for rehearing granted Jan. 17, 2012*.
[79] See description of this decision in text accompanying note 71 *supra*.
[80] 549 U.S. 497 (2011).
[81] *See* Michael Faure et al., CLIMATE CHANGE LIABILITY (2011); Richard Lord et al. (eds.), CLIMATE CHANGE LIABILITY: TRANSNATIONAL LAW AND PRACTICE (2012); Andrew L. Strauss, *The Legal Option: Suing the United States in International Forums for Global Warming Emissions*, 33 Envtl. L. Rptr. 10185 (2003).
[82] RESTATEMENT (THIRD) OF FOREIGN RELATIONS LAW §601(1). *See also* Legality of the Threat or Use of Nuclear Weapons, Advisory Opinion, 1996 ICJ Reports 226, 241-242 (July 8, 1996) ("the existence of the general obligation of states to ensure that activities within their jurisdiction and control respect the environment of other states or of areas beyond national control is now part of the corpus of international law relating to the environment").
[83] Trail Smelter (U.S. v. Canada), 3 R.I.A.A. 1938, 1965 (March 11, 1941).
[84] Petition to the Inter American Commission on Human Rights Seeking Relief from Violations Resulting from Global Warming Caused by Actions and Omissions of the United States, available at http://inuitcircumpolar.com/files/uploads/ icc-files/FINALPetitionICC.pdf.
[85] *Id.* at 5.
[86] See the discussion of the possible effects of climate change on water availability in the western United States in Priyanka Sundareshan, *Using the Transfer of Water Rights as a Climate Change Adaptation Strategy: Comparing the United States and Australia*, 27 Ariz. J. Int'l & Comp. L. 911, 920-921 (2010).
[87] Cal Const. art. 10, §2 (describing the principles of beneficial use and reasonableness as "self-executing"); State Water Resources Control Bd. v. Forni, 126 Cal. Rptr. 851 (1976) (noting that "[w]hat is a [reasonable and] beneficial use at one time may, because of changed conditions, become a waste of water at a later time").
[88] Sundareshan, *supra* note 86, at 923-925.
[89] Joel Smith et al., Georgetown Climate Center, *Adaptation Case Studies in the Western United States* at 22 (2011) (writing with specific reference to Colorado's prior appropriation doctrine).
[90] *See, e.g.*, Bassinger v. Taylor, 164 P. 522, 523 (Idaho 1917).
[91] Again using California as our example, that state's Supreme Court has explained that "overlying water rights are usufructuary only, and while conferring the legal right to use the water that is superior to all other users, confer no private right of ownership in public waters." City of Barstow v. Mohave Water Agency, 5 P.3d 853, 860 n.7 (2000). An illustrative decision on a takings challenge to a county restriction on withdrawal of

groundwater (not, so far as appears, for climate change reasons) is Allegretti & Co. v. County of Imperial, 42 Cal. Rptr. 3d 122 (Cal. App. 2006) (no physical or regulatory taking caused by 12,000 acre-feet per year limit imposed by county in groundwater withdrawal permit).

[92] Edwards Aquifer Auth. v. Day, 2012 WL 592729 *20 (Tex. Feb. 24, 2012).

[93] Tulare Lake Basin Water Storage Dist. v. United States, 49 Fed. Cl. 313 (2001); Klamath Irrigation Dist. v. United States, 635 F.3d 505 (Fed. Cir. 2011); Casitas Municipal Water Dist. v. United States, 2011 Westlaw 6017935 (Fed. Cl. Dec. 5, 2011).

[94] Stockton East Water Dist. v. United States, 101 Fed. Cl. 352 (2011).

[95] Casitas Municipal Water Dist. v. United States, 543 F.3d 1276 (Fed. Cir. 2008). On remand, the trial court found the claim unripe, 102 Fed. Cl. 443 (Fed. Cl. 2011), a determination now on appeal back to the Federal Circuit.

[96] As explained in the *Casitas* remand, 102 Fed. Cl. at 455, with reference to the state of California: Under the public trust doctrine, state agencies have the responsibility to protect trust resources associated with California's waterways, such as navigation, fisheries, recreation, ecological preservation, and related beneficial uses. .... Similarly, the reasonable use doctrine prohibits the waste, unreasonable use, unreasonable method of use, and unreasonable method of diversion of water. (citations omitted)

[97] In some locations, sea level relative to the adjacent land has "fallen" because the land has risen more than the sea level. Land may rise once relieved of the massive weight of retreating glaciers as the result of climate change, natural and human-induced. Cornelia Dean, *As Alaska Glaciers Melt, It's Land That's Rising*, New York Times, May 19, 2009, at A1.

[98] *See generally* James G. Titus, *Rising Seas, Coastal Erosion, and the Takings Clause: How to Save Wetlands and Beaches Without Hurting Property Owners*, 57 Md. L. Rev. 1279 (1998).

[99] Stop the Beach Renourishment, Inc. v.. Florida Dep't of Envtl. Prot., 130 S. Ct. 2592, 2598 (2010).

[100] *See, e.g.,* City of Long Branch v. Jui Yung Liu, 4 A.3d 542, 550 (N.J. 2010).

[101] Identifying the portion of coastal erosion attributable to sea level rise may be a challenge. One writer notes: "In many Gulf of Mexico states, ... the projected rate of beach loss due to sea level rise is overwhelmed by the current background rate of erosion." Donna M. Christie, *Sea Level Rise and Gulf Beaches: The Specter of Judicial Takings*, 26 J. Land Use & Envtl. L. 313, 314 (2011).

[102] *See* Joe Sax, *Some Unorthodox Thoughts About Rising Sea Levels, Beach Erosion and Property Rights*, 11 Vt. J. Envtl. L. 641, 645 (2010).

[103] J. Peter Byrne and Jessica Grannis, *Coastal Retreat Measures*, in Michael B. Gerrard and Katrina F. Kuh, (eds.), ADAPTATION TO CLIMATE CHANGE AND THE LAW: U.S. AND INTERNATIONAL ASPECTS (ABA, forthcoming 2012).

[104] 580 S.E.2d 116 (S.C. 2003). *See also City of Long Branch*, 4 A.3d at 550 ("[u]nder the common law, the owner of oceanfront property takes title to dry land added by accretion, but loses to the State title over land that becomes tidally flowed as a result of erosion"); Bollay v. California Office of Administrative Law, 122 Cal. Rptr. 3d 490, 493 (Cal. App. 2011) ("the mean high tide line may change over time, affecting the seaward boundary of property along the coast").

[105] Stop the Beach Renourishment, Inc. v. Florida Dep't of Envtl. Prot., 130 S. Ct. 2592, 2602 (2010) (emphasis in original). *See generally* Christie, *supra* note 101. As noted by Justice Kennedy in his *Stop the Beach* concurring opinion, the Due Process Clause also constrains

state courts from substantially reducing property rights by arbitrary or irrational decision. 130 S. Ct. at 2614-2617.

[106] 566 F.3d 490 (5th Cir. 2009).

[107] 2012 Westlaw 1059341 (Tex. Mar. 30, 2012).

[108] The Fifth Circuit recently remanded *Severance* to the district court for further proceedings on the Fourth Amendment unreasonable seizure claim consistent with the Texas Supreme Court's answers to the certified questions. 2012 Westlaw 1825179 (5th Cir. May 21, 2012).

[109] 38 Me. Rev. Stat. Ann. §§480-A through 480-HH.

[110] Dep't of Envtl. Prot. Admin. Code ch. 355, §5.C.

[111] *See, e.g.,* Strother v. City of Rockwall, 358 S.W.2d 462 (Tex. App. 2012) (taking claim based on redesignation of land as floodplain defeated by, among other reasons, fact that land continued to be used for rental).

[112] CRS Report RL34131, *Flood Damage Related to Army Corps of Engineers Projects: Selected Legal Issues*, by Cynthia Brougher.

[113] In re Katrina Canal Breaches Litigation, 2012 WL 678135 (5th Cir. Mar. 2, 2012).

[114] Nicholson v. United States, 77 Fed. Cl. 605 (2007) (United States' failure to adequately design, build, or maintain flood protection system in New Orleans before and after Hurricane Katrina did not effect taking; rather, property damage was due to flooding caused by storm surge and such flooding was not the direct, natural, or probable result of the flood protection system).

[115] *See generally* David M. Stein, *Flood of Litigation: Theories of Liability of Government Entities for Damages Resulting from Levee Breaches*, 52 Loy. L. Rev. 1335 (2006).

[116] Quebedeaux v. United States, No. 11-389L (Fed. Cl. filed Sept. 21, 2011); Big Oak Farms, Inc. v. United States, No. 11-275L (Fed. Cl. filed Sept. 9, 2011). *See generally* Brian Lee and Alice M. Noble-Allgire, *High Water in the Nation's Breadbasket: A Takings Analysis of the Government's Response to the Mississippi River's Great Flood of 2011*, 26 Probate and Property 28 (Jan./Feb. 2012).

[117] *See, e.g.,* United States v. Cress, 243 U.S. 316, 328 (1917); Arkansas Game & Fish Comm'n v. United States, 648 F.3d 1377, 1379 (Fed. Cir. 2011), *cert. granted*, 2012 Westlaw 1069212 (Apr. 2, 2012).

[118] The question whether flooding that is not "inevitably recurring" necessarily falls short of being a taking, even a temporary taking, is now before the Supreme Court. *Arkansas Game & Fish Comm'n, supra* note 116.

[119] *See* Report RL34131, *supra* note 112. The principal tort defenses, discussed in Report RL34131, are two—the Corps of Engineers typically asserting both in each case. First, there is §3 of the Flood Control Act of 1928, 33 U.S.C. §702c, declaring that "[n]o liability of any kind shall attach to or rest upon the United States for any damage from or by flood waters at any place ...." Second, there is the "discretionary function exemption" under the Federal Tort Claims Act (FTCA), under which no tort can be maintained against the United States if based on a federal official's "exercise or performance or the failure to exercise or perform a discretionary function." 28 U.S.C. §2680(a). This exemption from the FTCA waiver of sovereign immunity protects federal officials from liability for decisions where there is room for policy judgment and discretion, and would likely apply to Corps of Engineers decisions as to operation of the agency's facilities. By contrast, the waiver of sovereign immunity for Fifth Amendment takings claims against the United States, found in the Tucker Act (28 U.S.C. §1491), has no comparable exemptions.

[120] Darryl Fears, "Climate change fight intensifies in Virginia," Wash. Post Dec. 18, 2011, at A3.

[121] City of El Paso v. Ramirez, 349 S.W.3d 181 (Tex. App. 2011). *See generally* Annot., *Liability for overflow or escape of water from reservoir, ditch, or artificial pond*, 169 ALR 517.

[122] In a definitive study of possible regulatory adaptations to sea level rise, the following are listed as possible "regulatory tools": zoning and overlay zones, floodplain regulations, building codes and resilient design, setbacks/buffers, conditional development and exactions, rebuilding restrictions, subdivisions and cluster development, hard-armoring permits, soft-armoring permits, and rolling coastal management / rolling easement statutes. Jessica Grannis, Georgetown Climate Center, *Adaptation Tool Kit, Sea-Level Rise and Coastal Land Use: How Governments Can Use Land Use Practices to Adapt to Sea-Level Rise* (2011).

[123] In this report, "armoring" does not include levees erected for flood protection, though some writers would extend the term that far. Levees are treated separately in Section IV.D.

[124] Fla. Admin. Code R. 63B-33.002(5).

[125] For a fuller recitation of the takings arguments pro and con with respect to anti-armoring statutes, see J. Peter Byrne, *Rising Seas and Common Law Baselines: A Comment on Regulatory Takings Discourse Concerning Climate Change*, 11 Vt. J. Envtl. L. 625, 636-638 (2010).

[126] 580 S.E.2d 116 (S.C. 2003).

[127] United States (Lummi Nation) v. Milner, 583 F.3d 1174, 1189 (9th Cir. 2009), *citing* Revell v. People, 52 N.E. 1052, 1059 (Ill. 1898).

[128] *See generally* Wendy B. Davis, *Reasonable Use Has Become the Common Enemy*, 9 Alb. L. Envtl. Outlook J. 1, 9- 10 (2004).

[129] 583 F.3d 1174 (9th Cir. 2009).

[130] *Id.* at 1189-1190.

[131] *Id.* at 1190.

[132] Joe Sax, *Some Unorthodox Thoughts About Rising Sea Levels, Beach Erosion, and Property Rights*, 11 Vt. J. Envtl. L. 641, 642 n.7 (2010).

[133] Evan Lehmann, *Conservative lawmakers, protecting their beaches, also adapt to climate change* (E&E Climatewire Feb. 10, 2012).

[134] 130 S. Ct. 2592 (2011).

[135] *Id.* at 2611.

[136] Borough of Harvey Cedars v. Karan, 40 A.3d 75 (N.J. Super. App. Div.), *certification granted* (June 8, 2012).

[137] Fisher v. Town of Nags Head, 725 S.E.2d 99 (N.C. Ct. App. 2012).

[138] The opening paragraphs of this section draw their inspiration from *Coastal Retreat Measures, supra* note 103.

[139] Thus far, reactive retreat appears to be the more common, but the pattern may be shifting. For example, the Oregon Coastal Management Program recently recommended "using land-use planning processes to address climate change." Oregon Coastal Management Program, Department of Land Conservation and Development, *Climate Ready Communities: A Strategy for Adapting to the Impacts of Climate Change on the Oregon Coast* at 5 (Jan. 2009). And a Hawaii state representative has introduced legislation requiring her state and its counties to acknowledge climate change in any future development taking place on the islands. *Bill requires Hawaii to prepare for sea level rise* (E&E ClimateWire Jan. 26, 2012).

[140] As the text notes, in contrast with regulatory prohibitions the mere removal of government development incentives is unlikely to be held a taking. *See, e.g.*, Texas Landowners Rights Ass'n v, Harris, 453 F. Supp. 1025 (D.D.C. 1978), *aff'd mem.*, 598 F.2d 311 (D.C. Cir. 1979), in connection with the National Flood Insurance Program. Another incentive-

removing federal statute, the Coastal Barrier Resources Act, ended federal support (such as federal mortgage guarantees and federal flood insurance) for development on certain barrier islands. 16 U.S.C. §§3501-3510. It has generated no reported takings decisions.

[141] 505 U.S. 1003 (1992).

[142] *Id.* at 1021 n.10.

[143] *Id.* at 1029.

[144] *Id.* at 1031.

[145] *See, e.g.*, McQueen v. South Carolina Coastal Council, 580 S.E.2d 116, 119 n.5 (S.C. 2003).

[146] *See* F. Patrick Hubbard, *The Impact of Lucas on Coastal Development: Background Principles, the Public Trust Doctrine and Global Warming*, 16 Southeastern Envtl. L. J. 65, 80 (2007).

[147] One commentator would answer yes to both the footnoted text question, involving written notice, and the immediately preceding text questions, involving only constructive knowledge. He argues that "increasing awareness of [sea level rise] and its impacts as well as distribution of such information should inform analysis of coastal owners' RIBE in legal claims that government regulation or action has taken private property." Thomas Ruppert, *Reasonable Investment-Backed Expectations: Should Notice of Rising Seas Lead to Falling Expectations for Coastal Property Purchasers?*, 26 J. Land Use & Envtl. L. 239 (2011).

[148] *Id.* at 260, citing as an example Cal. Civ. Code §1103(v)(1)(A).

[149] *Id.* at 266-267.

[150] *See generally* James Wilkins, *Is Sea Level Rise "Foreseesable"? Does It Matter?*, 26 Vt. J. Envtl L. 437 (2011).

[151] 42 U.S.C. §§4001-4128.

[152] Special flood hazard areas are mapped by the Federal Emergency Management Agency, which administers the NFIP generally. 44 C.F.R. §59.2(b).

[153] *See, e.g.*, Adolph v. Federal Emergency Management Agency, 854 F.2d 732 (5th Cir. 1988); Gove v. Zoning Bd. Of Appeals, 831 N.E.2d 865, 871-875 (Mass. 2005); Responsible Citizens in Opposition to Floodplain Ordinance v. City of Asheville, 302 S.E.2d 204 (N.C. 1983). *But see* McDougal v. County of Imperial, 942 F.2d 668 (9th Cir. 1991) (fact that government's purpose in floodway designation was legitimate does not automatically preclude regulatory takings claim).

[154] *Adolph*, 854 F.2d 732 (holding that Federal Emergency Management Agency cannot be sued for taking based on parish's adoption of floodplain regulations to qualify for NFIP, because adoption was not federally coerced).

[155] This paragraph discussing disinvestment in public infrastructure was inspired by David Lewis, *Constitutional Property Law Analysis of State and Local Government Disinvestment in Infrastructure as a Coastal Adaptation Strategy* (2012) (student paper on file with author).

[156] *See, e.g.*, Jordan v. Canton, 265 A.2d 96 (Me. 1970).

[157] 16 U.S.C. §§3501-3510.

[158] *See, e.g.*, Jordan v. St. Johns County, 63 So. 3d 835 (Fla. App. 2011) (argument that county has so failed in its duty to maintain road as to deprive property owner of access states taking claim; government inaction in the face of an affirmative duty to act can support taking claim).

[159] "Amortization programs dovetail nicely with the traditional notion of land-use planning that nonconforming uses should be phased out gradually rather than terminated immediately." R. Meltz, D.H. Merriam, and R.M. Frank, THE TAKINGS ISSUE 433 (Island Press 1999). The value of an amortization period for avoiding takings is well-established. *See, e.g.*, Naegele Outdoor Advertising Co. v. City of Durham, 844 F.2d 172, 177 (4th Cir. 1988).

[160] 42 U.S.C. §§9601-9675.
[161] 40 C.F.R. §1506.11. See also the NEPA regulations of the Corps of Engineers, which call on that agency, in responding to emergencies, to refer actions with potentially significant environmental impacts to the CEQ as to NEPA arrangements "[w]hen possible." 33 C.F.R. §230.8.
[162] 42 U.S.C. §5159.
[163] 23 C.F.R. §771.117(d).
[164] Sierra Club v. Hassell, 636 F.2d 1095, 1099 (5th Cir. 1981) (replacement of bridge destroyed by hurricane requires no EIS). *Accord*, Citizens for the Scenic Severn River Bridge, Inc. v. Skinner, 802 F. Supp. 1325, 1333 (D. Md. 1991).
[165] 33 U.S.C. §1344(f)(1)(B).
[166] *See* Nationwide Permit No. 3 (repair, rehabilitation, or replacement of any previously authorized, currently serviceable structure), No. 31 (maintenance of existing flood control facilities), and No. 45 (restoration of upland areas damaged by storms, floods, or other discrete events, including bank stabilization). 77 Fed. Reg. 10,270 (Feb. 21, 2012).
[167] This section of the report was written by Ruth Wasem, Specialist in Immigration Policy, CRS Domestic Social Policy Division.
[168] "Conflicts Fuelled by Climate Change Causing New Refugee Crisis, Warns UN," by Julian Borger, *The Guardian*, (June 17, 2008), available online at: http://www.guardian.co.uk/environment/2008/jun/17/climatechange.food. See *also* United Nations High Commissioner on Refugees, THE STATE OF THE WORLD'S REFUGEES 2012: IN SEARCH OF SOLIDARITY ch. 7 ("Displacement, Climate Change, and Natural Disasters"), summary available at http://www.unhcr.org/publications/unhcr/sowr2012.
[169] The United States is not a party to the 1951 Convention but is a party to the 1967 Protocol Relating to the Status of Refugees, which amends the Convention. 19 U.S. Treaties 6223.
[170] *See* definition of "refugee" in INA §101(a)(42), 8 U.S.C. §1101(a)(42). This definition governs the reach of INA §207, 8 U.S.C. §1157, governing admissions based on humanitarian concerns, and INA §208, 8 U.S.C. §1158, governing asylum.
[171] For further background, see CRS Report RL31269, *Refugee Admissions and Resettlement Policy*, by Andorra Bruno; CRS Report R41753, *Asylum and "Credible Fear" Issues in U.S. Immigration Policy*, by Ruth Ellen Wasem; and CRS Report RS20844, *Temporary Protected Status: Current Immigration Policy and Issues*, by Ruth Ellen Wasem and Karma Ester.

Chapter 2

COMMON-LAW CLIMATE CHANGE LITIGATION AFTER *AMERICAN ELECTRIC POWER V. CONNECTICUT*[*]

Robert Meltz

SUMMARY

Congressional inaction on climate change has led concerned parties to explore other ways to address climate change—including lawsuits seeking to establish climate change impacts as a common law nuisance.

The prospects for these common law suits are limited, owing in part to the unsuitability of private litigation for dealing with global problems like climate change. Recently, the outlook for federal common-law suits seeking injunctive relief vis-a-vis climate change became particularly dim. On June 20, 2011, the Supreme Court ruled in *American Electric Power Co., Inc. v. Connecticut* that given EPA's Clean Air Act authority over greenhouse gas (GHG) emissions—affirmed by the Court a few years ago—the federal common law of nuisance in the area of climate change is "displaced." Federal courts may not use federal common law to add their own judge-made GHG emission standards to those of EPA.

The displacement of federal common law by *American Electric Power* is only one of three threshold issues that have bedeviled lawsuits seeking to establish climate change as a common law nuisance. The

---

[*] This is an edited, reformatted and augmented version of Congressional Research Service, Publication No. R41496, dated August 16, 2011.

standing inquiry requires a plaintiff in federal court to show actual or imminent injury caused by the defendant, and the likelihood that the injury will be redressed by the requested relief. Each of these factors can pose difficulties for the climate-change plaintiff. Similarly, the political question doctrine has led some courts to dismiss common-law climate change suits on the ground that the issue is better left with the political branches. Of course, where *American Electric Power* applies and the case must be dismissed on displacement grounds, standing and political question doctrine are now less important.

American Electric Power raises several questions. First, with federal common law displaced in the area of climate change, are state common law claims viable? Two threats to such claims are the possibility of preemption by the Clean Air Act (the sounder argument is against preemption), and the influence of the Supreme's Court's aversion to judge-made law in the climate change area so evident in *American Electric Power*. A second question is whether *American Electric Power* displaces climate-change-based federal common law actions when the remedy sought is monetary rather than injunctive. Finally, if Congress eliminates EPA authority over GHG emissions and is silent as to federal common law actions, does federal common law cease to be displaced so that such actions are again possible?

In addition to *American Electric Power*, there are two other active cases raising common law nuisance claims as to climate change—both involving coastal damage. In *Village of Kivalina v. ExxonMobil Corp.*, a coastal Eskimo village is suing energy companies alleging that their GHG emissions have contributed to shoreline erosion, requiring relocation of the village. In *Comer v. Murphy Oil*, Gulf coast landowners are suing energy and chemical companies asserting that their GHG emissions intensified Hurricane Katrina, adding to plaintiffs' property damage. Both of these cases raise the above-noted issue whether *American Electric Power* applies to actions seeking monetary damages.

A second common law theory recently has entered the fray. Since May 2011, either a suit or rulemaking petition has been filed in every state arguing that the respective state has a "public trust" duty to the atmosphere that requires it to address climate change. A suit has also been filed against the United States on the same ground.

I. INTRODUCTION

Congressional inaction on climate change has led various entities to pursue climate change measures off Capitol Hill. Either in hopes of making direct gains or to pressure Congress to act, such entities have looked to international forums, treaty negotiations, Environmental Protection Agency

(EPA) action under the Clean Air Act (CAA), state and regional efforts, and—the topic here—common law suits. The principal focus of such suits has been to establish greenhouse gas (GHG) emissions and climate change impacts as a nuisance.

For reasons discussed in this report, the prospects of this common law litigation are limited. Recently, the outlook for at least those cases based on the *federal* common law of nuisance and seeking injunctive relief has particularly dimmed. In 2007, the Supreme Court held in *Massachusetts v. EPA* that the CAA gives EPA authority to regulate GHG emissions from new motor vehicles (and, by implication, other GHG sources).[1] EPA responded by beginning to erect a regulatory edifice under that act for GHG emissions.[2] On June 20, 2011, the Supreme Court then delivered federal common law of nuisance suits a major blow. In *American Electric Power Co., Inc. v. Connecticut*, it held that in light of EPA's authority over GHG emissions as clarified in *Massachusetts*, federal common law in the climate change area is "displaced."[3] That is, federal courts may not use federal common law to add their own judge-made GHG emission standards, whether or not EPA exercises its authority. Thus, the potential of federal common-law climate change lawsuits seeking to have courts develop emission standards now seems poor.

Even before *American Electric Power*, many argued that courts should be unreceptive to dealing with a global problem as complex as climate change through individual common law suits, as opposed to a specifically tailored statute.[4] Each suit, after all, brings before the court only a handful of defendants representing a tiny fraction of the problem. As well, nuisance law offers no clear standards to apply. Questions of causation are also substantial: even if the court accepts that man-made GHG emissions contribute to climate change, how can a plaintiff show that a particular adverse impact was caused by climate change, and further was caused by GHG emissions of the defendants? And should the defendants' contribution to worldwide GHG emissions be viewed as *de minimis*—too small for a court to bother with? Questions of remedy are likely to be particularly intractable: what amount of emission reduction, or monetary compensation, should be required of a defendant given the likely miniscule fraction of worldwide GHG emissions contributed by that defendant?

Nonetheless, the use of nuisance lawsuits to attack climate change has its defenders.[5] They argue with some merit that even though nuisance law has never been used to deal with a problem as complex as climate change, many harms attributed to climate change—ecosystem and weather modifications, increased flooding, and harm to human health—are of a type traditionally covered by nuisance doctrine. And the Supreme Court has recognized that "public nuisance law, like common law generally, adapts to changing factual and scientific circumstances."[6]

By way of background, a nuisance may be either a private nuisance or a public nuisance. An activity constitutes a *private* nuisance if it is a substantial and unreasonable invasion of another's interest in the private use and enjoyment of land, without involving trespass.[7] Private nuisance actions are brought by the aggrieved landowner. An activity is a *public* nuisance if it creates an "unreasonable" interference with a right common to the general public.[8] Unreasonableness may rest on the activity significantly interfering with, among other things, public health and safety. Public nuisance cases are usually brought by the government rather than private entities, but may be brought by the latter if they suffer special injury.[9] Most of the common-law nuisance actions based on climate change have involved public nuisance.

Part II of this report notes the recurring threshold issues raised in nuisance litigation involving GHG emissions and climate change: "displacement" of federal common law, standing, and political question doctrine. By upholding the displacement barrier to suit, *American Electric Power* seems to have reduced the importance of the other two threshold issues. Part III describes the *American Electric Power* decision and speculates as to its likely aftermath and impact on climate change litigation generally. Part IV summarizes the other common law nuisance cases based on climate change, of which two remain active. Part V reviews the public trust doctrine suits, a recently filed group of cases that add a new common law theory to the litigation dealing with climate change.

II. RECURRING THRESHOLD ISSUES

As the court decisions in Parts III and IV show, the use of a nuisance action to address GHG emissions presents the plaintiff with daunting threshold hurdles—that is, issues that must be resolved at the outset of the litigation.[10] In light of *American Electric Power*, however, one of these threshold issues—whether the federal common law of nuisance has been displaced—will likely prove the key one in future efforts to use federal common law to address climate change. Thus, the role of the two other threshold issues, standing and political question doctrine, has been reduced.

A. Displacement of Federal Common Law

Because GHG emissions move across state lines, the federal rather than state common law of nuisance seems, at first blush, applicable. Though the Supreme Court barred federal courts from developing a "general" common law 73 years ago (they should instead apply the substantive law of the state in which they sit),[11] the Court has since clarified that in areas of national concern, such as interstate pollution, the articulation of federal common law by the federal courts is appropriate.[12]

But federal common law may be displaced by acts of Congress. Such judicially created law, says the Supreme Court, is a "necessary expedient," and "when Congress addresses a question previously governed by a decision rested on federal common law the need for such an unusual exercise of lawmaking by federal courts disappears."[13] Otherwise put, "new federal laws and new federal regulations may in time pre-empt the field of federal common law of nuisance."[14] Thus, the question arose early on in some of the climate change cases whether the federal CAA displaces judge-made law in the climate change area. As noted at the outset, the displacement argument was strengthened by the Supreme Court's 2007 decision in *Massachusetts v. EPA*, holding that EPA has CAA authority to regulate GHG emissions. With the Court's 2011 decision in *American Electric Power*, the displacement question has been resolved: as for any GHG source over which EPA has been delegated regulatory authority under the CAA, that statute eliminates any role for federal common law in abating GHG emissions. EPA using the CAA, not district court judges, will set GHG emission limits. The test for whether displacement has occurred, said the Court, is "whether the statute speaks directly to the question at issue."[15] Given the holding in *Massachusetts* and CAA coverage of existing stationary emission sources, the act definitely does "speak[] directly" to the defendants' GHG emissions. (See Section III.A. for a detailed description of what *American Electric Power* held.)

B. Other, Now Less Important, Threshold Issues

The threshold issues made less important by *American Electric Power* in federal-common-law climate change litigation are, again, the standing issue and political question doctrine. Their importance has not been eliminated, however, as there remains the possibility that *American Electric Power* will be held not to apply to all uses of federal common law in this area (see Section

III.B. as to actions seeking monetary damages), not to mention state common law cases.

The standing issue asks whether a party is an appropriate one to invoke the jurisdiction of a federal court created under Article III of the Constitution (this includes the district courts). Only a party with standing can bring suit in such courts. As developed by the Supreme Court, standing has constitutional and prudential (court-created) components. The constitutional side stems from the limitation of federal court jurisdiction in Article III to "Cases" and "Controversies." As explicated by the Court, this constraint demands that a plaintiff in federal court demonstrate (1) actual or imminent injury that is concrete and particularized, and not speculative; (2) that the injury is or will be caused by the defendant; and (3) that the injury likely will be redressed by a favorable court decision.[16]

A suit seeking relief from climate change impacts may run into difficulty with each of the three constitutional standing requirements. For example, climate change modeling generally predicts only large-scale effects, allowing defendants to argue in many cases that the particular injury suffered by plaintiff was not shown to have been caused by climate change. Or defendants might contend that their GHG emissions were (or will be) at best a *de minimis* contributor to plaintiff's injury.

State plaintiffs may have a choice. They may bring suit as owners of natural resources or other property, in which case they face the same standing requirements as private entities, described above. Alternatively, states may sue in their *parens patriae* capacity—that is, as protector of their quasi-sovereign interests—in which case the Article III requirement is differently stated. For *parens patriae* standing, a state must articulate a quasi-sovereign interest—that is, one apart from the interests of particular private parties. A state's interest in the "health and well-being—both physical and economic—of its residents in general,"[17] if a substantial portion of those residents is affected, is a well-established quasi-sovereign interest.[18] Owing to these quasi-sovereign interests, the Court said in *Massachusetts* in 2007 that states are "not normal litigants for purposes of invoking federal jurisdiction," but rather face a lower standing threshold.[19] Parenthetically, this was a 5-4 decision, and in *American Electric Power* in 2011 the Court's split on the standing issue was still evident—the holding of the court below that plaintiffs had standing was affirmed, but by an equally divided vote. The Court's even split (4-4) could happen, however, only because Justice Sotomayor recused herself; in a future case where she did not, the Court might vote 5-4 in favor of standing, or at least state standing, if the *American Electric Power* displacement barrier does not apply.

Unlike constitutional standing principles, the rules of prudential standing are not dictated by Article III. Rather, they are "judicially self-imposed limits on the exercise of federal jurisdiction."[20] One such prudential principle is "the rule barring adjudication of generalized grievances more appropriately addressed in the legislative branches."[21] Plainly this may be a concern with cases alleging climate change injuries, at least where such injuries are not concrete and personal.[22]

Political question doctrine leads a court to dismiss an action seen as presenting a "political question." The doctrine is "designed to restrain the Judiciary from inappropriate interference in the business of the other branches of Government."[23] However, deciding whether a matter has been committed by the Constitution to a nonjudicial branch of government is a "delicate exercise,"[24] and is decided on a case-by-case basis. The six factors indicating a non-justiciable political question were famously stated by the Supreme Court in *Baker v. Carr* in 1962.[25] Of these, the first three have played a role in the climate-change nuisance cases:

> Prominent on the surface of any case held to involve a political question is found [(1)] a textually demonstrable constitutional commitment of the issue to a coordinate political department; or [(2)] a lack of judicially discoverable and manageable standards for resolving it; or [(3)] the impossibility of deciding [the issue] without an initial policy determination of a kind clearly for nonjudicial discretion ...

Yet *Baker* made clear it was setting a high threshold for nonjusticiability. Since *Baker* was decided almost a half-century ago, the Court has found few issues to present political questions, but the doctrine has been ubiquitous in the nuisance/climate change litigation.

In *American Electric Power*, however, the Court was less than clear as to use of the political question doctrine in climate change litigation. The opinion remarks that "[f]our members of the Court would hold that at least some plaintiffs have Article III standing ... and further that *no other threshold obstacle bars review*."[26] The italicized phrase arguably includes the political question issue.

The opinion makes no comparable statement, however, as to the four other members of the Court; it notes only that they would find no standing.

III. AMERICAN ELECTRIC POWER

A. The Decision

American Electric Power originated when eight states, New York City, and three private land trusts brought nuisance actions, later consolidated, against five electric utility companies. The defendant utilities were chosen as allegedly the nation's largest emitters of CO_2, the major GHG, through their fossil-fuel electric power plants. Plaintiffs sought to require the electric utilities to abate their contribution to the nuisance of climate change by reducing their CO_2 emissions. No precise amount of emissions reduction was demanded. Plaintiffs cited both the federal common law of nuisance, and, in the alternative, state common law and statutory nuisance law.

In 2005, the federal district court dismissed the case on political question grounds. It held that because resolving the issues in the case required a balancing of economic, environmental, foreign policy, and national security interests, the court needed guidance from the political branches.[27] The absence of such guidance (there being no federal regulation of CO_2 as of 2005) meant to the court that the case satisfied one of the factors identified in Baker v. Carr as indicating a political question—namely, the case was "impossib[le] [to] decid[e] without an initial policy determination of a kind clearly for nonjudicial discretion." On appeal, the Second Circuit held in 2009 that the district court erred when it dismissed the case on political question grounds, that all plaintiffs had standing, and that the federal common law of nuisance had not been displaced by the CAA regulatory scheme.[28]

While all three threshold issues were presented to the Supreme Court in the petition for certiorari, the Court's decision was devoted almost entirely to the displacement question. This tight focus on displacement had been presaged by the oral argument before the Court, when nearly all the justices' questions were aimed in that direction—probably because displacement was the easiest-to-resolve threshold issue. The opening premise of the Court's opinion, which was unanimous on the displacement issue, was that when Congress addresses a question, "the need for such an unusual exercise of law-making [as federal common law] disappears."[29] "The test," it said, "for whether congressional legislation excludes ... federal common law is simply whether the statute speaks directly to the question at issue."[30]

So does the CAA "speak directly" to CO_2 emissions from existing fossil-fuel-fired power plants such as those of the defendants in the case? Yes, said the Court, owing to two simple facts. First, "Massachusetts made plain that emissions of carbon dioxide qualify as air pollution subject to regulation under the act."[31] Second, CAA section 111 instructs EPA to list categories of stationary sources that "contribute significantly to air pollution that may reasonably be anticipated to endanger public health or welfare"[32] and then establish standards of performance for new and modified sources in each category. Section 111(d) then requires regulation of existing sources within such categories—bringing in defendants' power plants. Concededly, 111(d) regulations are adopted by the states, but they are created pursuant to federal guidelines and receive federal oversight. Moreover, the fact that EPA has not yet actually exercised this authority as to GHG emissions from existing fossil-fuel-fired power plants is not the point. It is the delegation of the authority from Congress to EPA, the Court stressed, that displaces the common law, no matter how, or even whether, EPA chooses to exercise it. With this reasoning, the Court's holding was inescapable:

> The Clean Air Act and the EPA actions it authorizes displace any federal common law to seek abatement of carbon dioxide emissions from fossil-fuel fired [sic] power plants. *Massachusetts* [*v. EPA*] made plain that emissions of carbon dioxide qualify as air pollution subject to regulation under the Act.... And we think it equally plain that the Act "speaks directly" to emissions of carbon dioxide from the defendants' plants.

Buttressing its holding, the Court stressed the complex nature of climate change and the policy determinations on which government action must be based. "The Clean Air Act entrusts such complex balancing to EPA in the first instance,"[33] it said. "The expert agency is surely better equipped to do the job than individual district court judges issuing ad hoc, case-by-case injunctions."[34] In light of its holding, the Court remanded the case to the Second Circuit for further proceedings.

B. Likely Aftermath and Other Impacts

On the remand of *American Electric Power*, the Second Circuit presumably will dismiss the federal common law claims in the case. The court now may have to determine the fate of the *state* common law claims, which it did not address in its prior decision.[35] One issue will be whether the CAA preempts state common law claims regarding GHG emissions.[36] The answer might well be no, in light of CAA non-preemption provisions[37] and the general presumption against federal preemption of state law. Even if not preempted, however, plaintiffs asserting state common law may have an uphill climb. The Supreme Court's extended discussion in *American Electric Power* of why judges are ill-equipped to resolve climate change questions in the first instance (as opposed to during review of agency action) is likely to prove influential with courts adjudicating state as well as federal common law claims.

If the merits of the state common law nuisance claims are reached (in *American Electric Power* or other litigation), which state's common law will apply? Under relevant precedent, it is probable that the applicable state law will be that of the state where the particular GHG source is located— that is, a court probably will not apply the law of an affected state against an out-of-state source.[38] Applying the nuisance law of the source's state, a problem for plaintiffs may be establishing that a plant in compliance with state-issued permits can at the same time be a nuisance under that state's law.[39]

A provocative question now getting attention is whether *American Electric Power* displaces climate-change-based federal common law actions seeking *monetary* relief.[40] In that case, plaintiffs sought only *injunctive* relief: a court order requiring the defendants to reduce their GHG emissions. The Court's reasons for finding displacement seem heavily skewed to that form of relief—for example, the lack of federal court expertise for setting GHG emission standards, and the unacceptability of having EPA standards and judicial standards as parallel tracks. These concerns are arguably not present in a monetary damages case where the court's only task is to determine, by a preponderance of the evidence, that plaintiff's injury was proximately caused by the defendant's emissions. No standard setting is involved. As the next section notes, the applicability of *American Electric Power* to cases seeking damages may be resolved soon in *Village of Kivalina*.

Finally, the Supreme Court decision has no direct effect on EPA's emerging GHG regulation program. Indirectly, however, the decision gives the program added impetus. For one thing, it reaffirms the *Massachusetts v. EPA* holding that the CAA authorizes EPA to regulate GHG emissions. For another, it underscores the complexity of climate change and the consequent need for administrative expertise such as EPA's in grappling with it. Of course, if Congress succeeds in eliminating EPA authority over GHG emissions, or certain sources of such emissions, a very different question arises. In the (perhaps unlikely) event that such a law would be silent as to its intended impact on common law claims, it could be argued that elimination of EPA authority over GHGs also eliminates any displacement of federal common law. That resurrects the possibility of judge-made emission standards, if it is determined that climate change constitutes a nuisance.

IV. OTHER COMMON LAW OF NUISANCE CASES BASED ON CLIMATE CHANGE

Three climate change cases invoking the common law of nuisance are currently active. One is *American Electric Power Co.*, described above. The others are *Village of Kivalina v. ExxonMobil Corp.* and *Comer v. Murphy Oil USA*, discussed here. None of these pending cases, nor the finally resolved cases discussed afterward, have seen anything approaching a decision on the merits—all have been preoccupied exclusively with threshold issues. Thus we do not yet know whether GHG emissions can constitute a nuisance.

In *Village of Kivalina*, an Inupiat Eskimo village on the northwest Alaska coast sued 24 oil and energy companies, claiming that the large quantities of GHGs they emit contribute to climate change. Climate change, the village contends, is destroying the village by melting Arctic sea ice that formerly protected it from winter storms, leading to massive coastal erosion that will require relocating the village's inhabitants at a cost of $95 million to $400 million. Plaintiffs invoke the federal common law of public nuisance, and state statutory or common law of private and public nuisance. They further press a civil conspiracy claim, asserting that some of the defendants have engaged in agreements to participate in the intentional creation or maintenance of a public nuisance—that is, global warming—by misleading the public as to the science of global warming. The suit seeks monetary damages.

In 2009, the district court held that the federal nuisance claim was barred by political question doctrine and lack of standing.[41] The village appealed to the Ninth Circuit,[42] which stayed the case pending the Supreme Court decision in *American Electric Power*. The reactivation of this case is the first judicial development following that decision. Counsel for plaintiffs reportedly are arguing that *American Electric Power* applies only to injunctive-relief cases, not, as here, where a monetary remedy is sought. Note also in the preceding paragraph that there are claims in this case other than those based on federal common law.

As an aside, the liability insurer of one of the *Kivalina* defendants has filed suit seeking a declaratory judgment that should the defendant be found liable for damages in *Kivalina*, the insurer's general liability policies with the defendant will not apply.[43]

Comer v. Murphy Oil USA litigation has been reactivated after a seeming demise. Owners of Gulf coast property damaged by Hurricane Katrina sued certain oil, coal, and chemical companies under state law. They alleged a multistep chain of causation—that the GHGs emitted by the defendant companies, by contributing to global warming with consequent sea level rise and warmer sea water, caused Hurricane Katrina to intensify and increased the harm to plaintiffs' property. On this basis, plaintiffs asserted state-law tort claims, including negligence, nuisance (public and private), and trespass, and sought compensatory damages. They also requested punitive damages for gross negligence. Further, they claimed conspiracy to commit fraudulent misrepresentation, alleging, as in *Village of Kivalina*, that the oil and coal companies disseminated misinformation about global warming. Finally, plaintiffs made claims against their home insurance companies (e.g., breach of fiduciary duty claim for misrepresenting policy coverage, and violation of a state consumer-protection act) and their mortgage companies (arguing that they may not claim sums owed by plaintiffs for the value of the mortgaged property that was uninsured).

The federal district court dismissed the action for lack of plaintiff standing, and also found the claims precluded by the political question doctrine.[44] Then, in 2009, the Fifth Circuit reversed.[45] Relying on the Supreme Court's approval of standing in *Massachusetts v. EPA*, the panel ruled that the *Comer* plaintiffs similarly had Article III standing as to their tort claims. Plaintiffs, however, were held to lack standing as to their other claims. On the other major issue in the case, the circuit court held, contrary to the district court, that the tort claims were not barred by the political question doctrine. At this point, however, events took an odd turn. In 2010, after taking the case *en banc*, the Fifth Circuit announced it lacked a quorum, so the appeal had to be dismissed.[46] Indeed, the court concluded it could not even reinstate the vacated panel decision. The effect was to deny appeal of the original district court dismissal, which the Fifth Circuit effectively reinstated. However, on May 27, 2011, the plaintiffs refiled the case (with minor modifications), creating a second opportunity for a ruling on whether *American Electric Power* applies to cases seeking damages.

The two no-longer-active cases deserve but brief mention. In *California v. General Motors Corp.*, that state sued auto manufacturers based on the alleged contributions of their vehicles, through GHG emissions, to climate change impacts in the state. The suit asserted that these impacts constitute a public nuisance under federal common law, and sought damages. In 2007, the district court dismissed on a political question rationale.[47] California appealed to the Ninth Circuit, but in 2009 motioned for voluntary dismissal, which the circuit granted. Dismissal was sought as part of an agreement between the state, the Obama Administration, and the automobile manufacturers. Finally, *Korsinsky v. U.S. EPA* was a *pro se* action apparently alleging that GHG emissions, by contributing to climate change, threatened plaintiff's health due to his enhanced vulnerability as an older person with sinus problems. He appeared to have requested an injunction ordering EPA to require less pollution and ordering polluters to use his invention for reducing CO_2 emissions. The district court dismissed for lack of standing, and the Second Circuit affirmed on the same ground in 2006.[48]

V. A NEW COMMON LAW THEORY ENTERS THE FRAY: PUBLIC TRUST DOCTRINE

Since May 2011, the nuisance lawsuits above have been joined by a coordinated campaign of lawsuits and rulemaking petitions seeking to attack climate change by an entirely different common law theory: public trust doctrine. The claim is that the states and the federal government have a public trust responsibility to protect the atmosphere, and have failed to exercise that responsibility to deal with the threat of climate change. Many of the plaintiffs and petitioners are children and teenagers, represented by their guardians ad litem. The lawsuits and petitions are being coordinated by Our Children's Trust, an Oregon nonprofit.[49]

As background, the public trust doctrine is an ancient common law principle with origins in Roman law and the Magna Carta. It asserts that certain natural resources are held by the sovereign in special status.[50] Key aspects of that special status are that government may neither alienate public trust resources nor, more pertinent here, permit their injury by private parties. Rather, government has an affirmative duty to safeguard these resources for the benefit of the general public. The doctrine is generally a principle of state law, though there is limited recognition of a federal counterpart. After tidelands and the beds of navigable waterways, fish and wildlife are the natural resources most traditionally associated with the public trust doctrine; courts do not appear to have applied the doctrine to the atmosphere yet, as the suits and petitions here are seeking.

As for the lawsuits, each one reportedly asks the court for declaratory relief proclaiming that the atmosphere is a public trust resource and that the government in question has a fiduciary duty as trustee to protect it. Twelve suits have been filed—against the United States, Alaska, Arizona, California, Colorado, Iowa, Minnesota, Montana, New Mexico, Oregon, Texas, and Washington.[51] The Montana suit is unique in alleging a basis for extending the public trust to the atmosphere under the state constitution and state statute.[52] Some of the suits ask for injunctive relief as well. For example, the suit against the United States asserts that the federal government has violated its trustee duties by allowing unsafe amounts of GHGs into the atmosphere and asks for an injunction requiring it to take action "consistent with the United States government's equitable share of the global effort."[53] None of the suits seek monetary damages.

The rulemaking petitions cover each state where no lawsuit was filed.[54] Each one, CRS is informed, cites the public trust doctrine and asks the appropriate state agency to regulate GHG emissions based thereon.[55] At this writing, a few of the petitions have been dismissed.

End Notes

[1] 549 U.S. 497 (2007).

[2] 74 Fed. Reg. 66,496 (Dec. 15, 2009) (EPA finalizes endangerment finding for GHG emissions from new motor vehicles); 75 Fed. Reg. 25,323 (May 7, 2010) (EPA GHG emission standards for new light-duty motor vehicles); and 75 Fed. Reg. 31,514 (June 3, 2010) (EPA promulgates "tailoring rule" limiting new source review of stationary sources of GHG emissions and limiting Title V permitting requirements). Additionally, on December 21, 2010, EPA entered into a settlement in which it agreed to issue new source performance standards for GHG emissions from electric power plants and oil refineries by 2012.

[3] 2011Westlaw 2437011 (U.S. June 20, 2011) (No. 10-174).

[4] *See, e.g.*, Daniel A. Farber, *Basic Compensation for Victims of Climate Change*, 155 U. Pa. L. Rev. 1605, 1649 (2007) ("Realistically, the greatest function of litigation may be to prod legislative action."). *See also* Jim Gitzlaff, *Getting Back to Basics: Why Nuisance Claims Are of Limited Value in Shifting the Costs of Climate Change*, 39 Envtl. L. Rptr. 10,218 (March 2009). For a judicial take on the common law versus statute question, see *North Carolina v. Tennessee Valley Auth.*, 615 F.3d 291, 302 (4th Cir. 2010) ("The contrast between the defined standards of the Clean Air Act and an ill-defined omnibus tort of last resort could not be more stark.").

[5] *See, e.g.*, Randall S. Abate, *Public Nuisance for the Environmental Justice Movement: The Right Thing and the Right Time*, 85 Wash. L. Rev. 197 (2010); Matthew F. Pawa, *Global Warming: The Ultimate Public Nuisance*, 39 Envtl. L. Rptr. 10,230 (March 2009); Jonathan Zasloff, *The Judicial Carbon Tax: Reconstructing Public Nuisance and Climate Change*, 55 UCLA L. Rev. 1827 (2008) (arguing that a nuisance-based climate change regime

essentially becomes a carbon tax); Daniel V. Mumford, *Curbing Carbon Dioxide Emissions Through the Rebirth of Public Nuisance Laws— Environmental Legislation by the Courts*, 30 Wm. & Mary Envtl. L. & Policy Rev. 195 (2005); David A. Grossman, *Warming Up to a Not-So-Radical Idea: Tort-Based Climate Change Litigation*, 28 Colum. J. Envtl. L. 1 (2003).

[6] *American Elec. Power*, 2011 Westlaw 2437011, *8.
[7] RESTATEMENT (SECOND) OF TORTS § 821D (1979).
[8] *Id.* at § 821B.
[9] To have suffered "special injury," a person must have incurred a different kind of interference than that suffered by the public at large, not just a greater harm from the same kind of interference. *Id.* at § 821B comments b. and d.
[10] *See generally* Kevin A. Gaynor et al., *Challenges Plaintiffs Face in Litigating Federal Common Law Climate Change Claims*, 40 Envtl. L. Rptr. (News and Analysis) 845 (Sept. 2010); Thomas W. Merrill, *Global Warming as a Public Nuisance*, 30 Colum. J. Envtl. L. 293 (2005).
[11] Erie R. Co. v. Tompkins, 304 U.S. 64, 78 (1938), *quoted with approval in American Elec. Power*, 2011 Westlaw 2437011 at *7.
[12] *See, e.g.,* Illinois v. Milwaukee, 406 U.S. 91, 103 (1972) ("Milwaukee I") ("When we deal with air and water in their ambient or interstate aspects, there is a federal common law."), *quoted with approval in American Elec. Power*, 2011 Westlaw 2437011 at *7.
[13] Milwaukee v. Illinois, 451 U.S. 304, 314 (1980) ("Milwaukee II"), *quoted with approval in American Elec. Power.*, 2011 Westlaw 2437011 at *9.
[14] *Milwaukee I*, 406 U.S. at 107. Indeed, there is a presumption in favor of such preemption. Matter of Oswego Barge Corp., 664 F.2d 327, 335 (2d Cir. 1981).
[15] American Elec. Power Co., Inc. v. Connecticut, 2011 Westlaw 2437011, *9 (June 20, 2011) (brackets and quotation marks omitted).
[16] *See, e.g.,* Lujan v. Defenders of Wildlife, 504 U.S. 555, 560-561 (1992).
[17] Alfred L. Snapp & Son, Inc. v. Puerto Rico, 458 U.S. 592, 607 (1982).
[18] *Id.* at 604-605.
[19] Massachusetts v. EPA, 549 U.S. 497, 518 (2007).
[20] Elk Grove Unified School Dist. v. Newdow, 542 U.S. 1, 11 (2004), *quoting* Allen v. Wright, 468 U.S. 737, 751 (1984).
[21] *Elk Grove Unified School Dist.*, 542 U.S. at 12.
[22] The Supreme Court has expressly rejected the argument that just because climate change causes widespread harm, standing doctrine presents an insurmountable obstacle to establishing federal jurisdiction. But "[w]hile it does not matter how many persons have been injured by the alleged action [being challenged], the party must show that the action injures him in a concrete and personal way." Massachusetts v. EPA, 549 U.S. 497, 517 (2007), *quoting* Lujan v. Defenders of Wildlife, 504 U.S. 555, 581 (1992) (Kennedy, J., concurring).
[23] United States v. Munoz-Flores, 495 U.S. 385, 394 (1990).
[24] *Baker*, 369 U.S. at 211.
[25] 369 U.S. 186, 216 (1962).
[26] American Elec. Power Co., Inc. v. Connecticut, 2011 Westlaw 2437011, *7 (June 20, 2011) (U.S. No. 10-174) (emphasis added).
[27] 406 F. Supp. 2d 265 (S.D.N.Y. 2005).
[28] 582 F.3d 309 (2d Cir. 2009).
[29] 2011 Westlaw 2437011, *9.

[30] *Id.* (emphasis added).
[31] *Id.*
[32] CAA § 111(b)(1)(A); 42 U.S.C. § 7411(b)(1)(A).
[33] 2011 Westlaw 2437011, *11.
[34] *Id.*
[35] Since the Second Circuit had found that the federal common law claims in the case were viable, those claims precluded any role for state common law. *See* International Paper Co. v. Ouellette, 479 U.S. 481, 488 (1987). Thus, the court did not have to adjudicate the plaintiffs' state common law claims.
[36] The Supreme Court expressly reserved this question.
[37] *See* CAA § 304(e), 42 U.S.C. § 7604(e) (providing that nothing in the CAA citizen suit section "shall restrict any right which any person ... may have under any statute or common law to seek enforcement of any emission standard or limitation or to seek any other relief...."); CAA § 116, 42 U.S.C. § 7416 ("... nothing in this act shall preclude or deny the right of any State or political subdivision thereof to adopt or enforce" any air pollution standard or requirement).
[38] This principle, that the law of the source state governs, was announced in International Paper Co. v. Ouellette, 479 U.S. 481, 492-494 (1987). *International Paper* arose in the context of interstate *water* pollution. In a recent case, however, the court emphatically held that the same principle applies to interstate *air* pollution. North Carolina v. Tennessee Valley Auth., 615 F.3d 291, 306 (4th Cir. 2010).
[39] *North Carolina*, 615 F.3d at 309 ("It would be odd, to say the least, for specific state laws and regulations to expressly permit a power plant to operate and then have a generic statute countermand those permissions on public nuisance grounds.").
[40] *See, e.g.,* Michael B. Gerrard, *'American Electric Power' Leaves Open Many Questions for Climate Litigation*, New York Law Journal, July 14, 2011, and discussion of *Village of Kivalina v. ExxonMobil* in Part IV of report.
[41] 663 F. Supp. 2d 863 (N.D. Cal. 2009).
[42] No. 09-17490.
[43] Steadfast Ins. Co. v. The AES Corp., No. 08-858 (Arlington County, Va., Cir. Ct. filed July 2008).
[44] 2007 WL 6942285 (S.D. Miss. August 30, 2007).
[45] 585 F.3d 855 (5th Cir. 2009).
[46] 607 F.3d 1049 (5th Cir. 2010).
[47] 2007 WL 2726871 (N.D. Cal. September 17, 2007).
[48] 192 Fed. Appx. 71 (2d Cir. 2006).
[49] Further details are available at http://www.ourchildrenstrust.org.
[50] *See generally* Joseph Sax, *The Public Trust Doctrine in Natural Resource Law: Effective Judicial Intervention*, 69 Mich. L. Rev. 471 (1970), and Jan S. Stevens, *The Public Trust: A Sovereign's Ancient Prerogative Becomes the People's Environmental Right*, 14 U.C. Davis L. Rev. 195 (1980).
[51] Links to the filed complaints may be found at http://www.ourchildrens trust.org.
[52] Barhaugh v. State of Montana, No. OP 11-0258 (Mont. Supreme Ct. filed May 4, 2011). The complaint cites, in the Montana constitution, art. IX, sec. 1 ("The state ... shall maintain ... a clean and healthful environment in Montana for present and future generations."), and art. II, sec. 3 ("All persons are born free and have certain inalienable rights. They include the right to a clean and healthful environment....").
[53] Alec L. v. Jackson, No. 11-CV-2203 (N.D. Cal. filed May 4, 2011).

[54] *See* http://www.ourchildrenstrust.org.
[55] Telephone conversation with Julia Olson, Executive Program Program Director, Our Children's Trust. Ms. Olson is the senior attorney coordinating the lawsuits and petitions.

In: Climate Change: Legal Issues and Contexts ISBN: 978-1-62257-847-4
Editors: B. Saunders and R. A. Diaz © 2013 Nova Science Publishers, Inc.

Chapter 3

FEDERAL AGENCY ACTIONS FOLLOWING THE SUPREME COURT'S CLIMATE CHANGE DECISION IN *MASSACHUSETTS V. EPA*: A CHRONOLOGY[*]

Robert Meltz

SUMMARY

On April 2, 2007, the Supreme Court rendered one of its most important environmental decisions. In *Massachusetts v. EPA*, the Court held 5-4 that greenhouse gases (GHGs), widely viewed as contributing to climate change, constitute "air pollutants" as that phrase is used in the Clean Air Act (CAA). As a result, said the Court, the U.S. Environmental Protection Agency (EPA) had improperly denied a petition seeking CAA regulation of GHG emissions from new motor vehicles by saying the agency lacked authority over such emissions.

This report offers a chronology of major federal agency actions, mainly by EPA, that involve GHGs or climate change and that occurred in the wake of *Massachusetts v. EPA*. Most of the listed actions trace directly or indirectly back to the decision. Examples include EPA's "endangerment finding" for GHG emissions from new motor vehicles; the agency's standards for GHG emissions from new motor vehicles; its

---

[*] This is an edited, reformatted and augmented version of Congressional Research Service, Publication No. R41103, dated May 1, 2012.

interpretation of "pollutants subject to regulation," the CAA trigger for requiring best available control technology (BACT) for such pollutants in "prevention of significant deterioration" areas; its guidance for determining BACT for GHG emissions; the "tailoring rule" (limiting the stationary sources that initially will have to install BACT and obtain CAA Title V permits based on their GHG emissions); and settlements of litigation seeking to compel new source performance standards (NSPSs) for GHG emissions from electric power plants and petroleum refineries. A few agency actions were included solely because of their relevance to climate change and their post-*Massachusetts* occurrence—for example, EPA's responses to California's request for a waiver of CAA preemption allowing that state to set its own limits for GHG emissions from new motor vehicles, and EPA's monitoring rule for GHG emissions.

INTRODUCTION

On April 2, 2007, the Supreme Court rendered one of its most important environmental decisions ever. The case, *Massachusetts v. EPA*,[1] arose when the U.S. Environmental Protection Agency (EPA) denied a petition asking it to take two actions—(a) find under the Clean Air Act (CAA) that greenhouse gases (GHGs) emitted from new motor vehicles "cause, or contribute to, air pollution which may reasonably be anticipated to endanger public health or welfare,"[2] through their climate change effects, then (b) issue standards for those GHG emissions. EPA's petition denial was based in part on its claim that it lacked authority to regulate GHGs. To the contrary, said the Supreme Court by 5-4, GHGs constitute "air pollutants" under the CAA, hence EPA does indeed have the authority to regulate GHG emissions. The Court gave EPA three options: (a) determine that GHG emissions from new motor vehicles "cause, or contribute to, air pollution which may reasonably be anticipated to endanger public health or welfare"; (b) determine that such GHG emissions do *not* do so; or (c) explain why the agency is unable to make a determination under either (a) or (b).

EPA chose option (a)—that is, to make a positive "endangerment finding" for GHG emissions from new motor vehicles. That finding was made in December 2009, whereupon the CAA required EPA to promulgate standards to address the endangerment.[3] That action the agency also has taken, in May 2010. These standards in turn triggered the CAA requirement that major new stationary sources and major modifications of existing stationary sources, when proposed for Prevention of Significant Deterioration (PSD) areas, must install "best available control technology" (BACT) to control GHG emissions.[4] And that requirement—to install BACT— triggered CAA Title V permitting requirements that for many air pollution sources would not otherwise be triggered.[5]

This report is a chronology of the major climate-change-related actions taken by federal agencies, principally EPA, in the wake of *Massachusetts v. EPA*. Most of the listed actions trace directly or indirectly back to the decision; a few were included solely because of their relevance to climate change and their occurrence post-*Massachusetts*. More analytical treatment of the agency actions in this report may be found in other CRS reports.[6]

The EPA actions listed in this report may be short-lived. Congressional efforts to delay or bar EPA regulation of GHGs while Congress is deliberating a post-CAA climate change regime have seen their prospects improve as a result of the 2010 election. For the time being, however, EPA's efforts to control GHG emissions under its CAA authority have been termed "the only game in town."[7] The potential impacts of EPA's efforts have been lost on no one: no fewer than 90 petitions for review of EPA's actions in this report have been filed in the U.S. Court of Appeals for the D.C. Circuit, all of them now pending.[8]

- A table of acronyms is provided in the **Appendix**.
- Dates used are those of Federal Register publication wherever a Federal Register citation is given. In most cases, however, the agency action was signed and publicly announced weeks earlier.
- Once an agency promulgates a final rule, the entry for the proposed rule has been deleted.

2008

March 6: EPA denies California's request for waiver of CAA preemption. 73 Fed. Reg. 12,156. By way of background, the CAA preempts state controls on new motor vehicle emissions,[9] but offers California, and California alone, the opportunity to request a waiver of CAA preemption.[10] EPA must grant the preemption waiver if certain conditions are met.[11] The importance of this "California waiver" is magnified by the fact that once EPA grants the waiver, states that adopt motor vehicle emission standards identical to California's also partake of the preemption waiver for the same vehicles.[12] In the present case, California sought a waiver of CAA preemption for its

GHG emissions limits for 2009 and later model year motor vehicles. EPA denied the waiver on finding that the state did not need those emission limits to meet "compelling and extraordinary conditions," as required by the CAA.[13] (See "July 8, 2009" below for EPA's reversal of this denial.)

July 30: EPA issues advance notice of proposed rulemaking. 73 Fed. Reg. 44,354. This document, titled "Regulating Greenhouse Gas Emissions Under the Clean Air Act," sets out EPA's view of the legal implications were EPA to make a positive endangerment finding for GHGs from new motor vehicles—as discussed in the "Introduction," option (a) offered by the Supreme Court. It is purely an informational document, prepared after the George W. Bush Administration decided in late 2007 not to issue an endangerment finding for new motor vehicle GHG emissions, but rather to leave that decision to the next Administration.

December 31: EPA Administrator publishes interpretive memorandum ("Johnson memorandum"). 73 Fed. Reg. 80,300. EPA Administrator Stephen Johnson issued this memorandum, titled "EPA's Interpretation of Regulations that Determine Pollutants Covered by Federal Prevention of Significant Deterioration (PSD) Permit Program." The memorandum narrowly interprets the CAA phrase "pollutant subject to regulation under this act"[14] to include only pollutants regulated by *actual, not potential future*, emission limits under the CAA or its regulations.

Much hangs on this distinction between actual, and potential future, emission limits. In PSD areas of the country, the CAA requires only pollutants "subject to regulation under [the CAA]" to be controlled by potentially expensive BACT—when emitted by new major emitting facilities or major modifications of existing facilities. Since there were no "actual" GHG regulations under the CAA when the Johnson memorandum was issued, this meant that for the near term at least, new major emitting facilities and major modifications of existing facilities proposed for PSD areas did not have to install BACT for GHG emissions.

2009

February 17: EPA grants petition for reconsideration of Johnson memorandum. (See "December 31, 2008" above.) EPA did not grant a stay of the memorandum, however, announcing that it will remain in effect until the agency makes a final decision at the end of the reconsideration period.

July 8: EPA grants California's request for waiver of CAA preemption. 74 Fed. Reg. 32744. This rule reversed EPA's prior denial of California's request for a preemption waiver (see "March 6, 2008" above). As noted, its effect is to allow California's GHG emissions limits for 2009 and later model year motor vehicles to go into effect, and to allow the identical emission standards for the same vehicles promulgated by other states to do likewise. Such "other states" now number 13, plus the District of Columbia.

October 7: EPA implements the grant of reconsideration of the Johnson memorandum. 71 Fed. Reg. 51,535. This document discusses various possible interpretations of "subject to regulation" and requests public comment. The interpretations discussed include EPA's "current and preferred interpretation, which would make PSD applicable to a pollutant on the basis of an EPA regulation requiring actual control of emissions of a pollutant." *Id.* at 51,535. (See "February 17, 2009" above.)

October 30: EPA finalizes mandatory GHG monitoring rule. 74 Fed. Reg. 56,260. The FY2008 Consolidated Appropriations Act[15] requires that EPA develop a rule "to require mandatory reporting of GHG emissions above appropriate thresholds in all sectors of the economy"—using EPA's existing CAA authority. The rule took effect January 1, 2010, with the first monitoring reports due in 2011. To allow additional time for setting up the reporting system, EPA recently extended the reporting deadline for 2010 emissions from March 31, 2011, to September 30, 2011. 76 Fed. Reg. 14,812 (March 18, 2011). Corrections and clarifications of the October 30, 2009, monitoring rule appear at 75 Fed. Reg. 66,434 (October 28, 2010). With the foregoing exceptions, this CRS report does not list EPA's many amendments and expansions of this monitoring rule. To stay abreast, the reader is referred to http://www.epa.gov/climatechange/ emissions/ghgrulemaking.html.

December 15: EPA finalizes endangerment finding for GHG emissions from new motor vehicles. 74 Fed. Reg. 66,496. This action under CAA Section 202(a)[16] was option (a) offered to EPA by the Supreme Court decision, as described on page 1. By this endangerment finding, EPA actually makes two findings under Section 202(a): first, that six GHGs currently in the atmosphere are reasonably likely to endanger both public health and welfare, and second, that the four GHGs emitted by new motor vehicle emissions in the United States contribute to that air pollution. The endangerment finding has no effect on outside parties *in itself*; its importance is that it triggers a duty under Section 202(a) for EPA to promulgate emission standards for the source category creating the endangerment—in this case, new motor vehicles. (See "May 7, 2010" below for emission standards.)

2010

February 8: Securities and Exchange Commission (SEC) issues guidance regarding corporate disclosure related to climate change. 75 Fed. Reg. 6290. This interpretive release provides guidance to public companies as to how existing SEC disclosure requirements apply to climate change matters.

February 18: Council on Environmental Quality (CEQ) issues draft guidance under National Environmental Policy Act (NEPA).[17] This guidance memorandum from CEQ is titled "Draft NEPA Guidance on Consideration of the Effects of Climate Change and Greenhouse Gas Emissions." It addresses the ways in which federal agencies can improve their consideration of GHG effects in their evaluation of proposals for federal actions under NEPA, including in environmental impact statements.

April 2: EPA finalizes its reconsideration of Johnson memorandum. 75 Fed. Reg. 17,004. After taking comments on alternate interpretations of "subject to regulation" (see "December 31, 2008" above), EPA decided to continue with the interpretation published December 31, 2008, in the Johnson memorandum (more recently referred to as the "timing rule"). In a refinement, however, EPA stated that "subject to regulation" does not apply to a newly regulated pollutant (like GHGs) until a regulatory requirement to control emissions of that pollutant not only is promulgated, but also *takes effect*. For GHGs, that "regulatory requirement" is the new GHG emission standards for light-duty motor vehicles, noted immediately below. Since these standards do not take effect until January 2, 2011, PSD and Title V permitting requirements also will not go into effect until then—or later under EPA's tailoring rule finalized June 3, 2010.

May 7: EPA and NHTSA jointly finalize rules setting GHG emission standards and fuel economy standards for 2012-2016 model year light-duty vehicles. 75 Fed. Reg. 25,323. The EPA emission standards (known as the "tailpipe rule") are pursuant to the agency's mandatory CAA duty to promulgate such standards once it finalizes its "endangerment finding" for new motor vehicles (see "December 15, 2009" above). Regarding NHTSA, the Energy Policy and Conservation Act, as amended in 2007, requires that agency to prescribe separate fuel economy standards for passenger and non-passenger automobiles beginning with model year 2011, to achieve a combined fuel economy average for model year 2020 of at least 35 miles per gallon.[18] EPA and NHTSA acted jointly because motor vehicle GHG emissions are directly linked to fuel consumption. In order to provide a consistent set of standards for auto manufacturers to meet, the White House brokered an agreement under which EPA would develop GHG emissions standards under the CAA that would be compatible with fuel economy standards developed by NHTSA.

The EPA and NHTSA standards apply to passenger cars, light-duty trucks, and medium-duty passenger vehicles, covering model years 2012 through 2016, and purport to represent a harmonized and consistent national program. (California has announced its commitment to support the national program: on April 1, 2010, it revised its GHG standards for model years 2012-2016 such that compliance with the federal GHG standards will be deemed compliant with California's GHG standards.[19]) Both EPA and NHTSA standards become more stringent each year, culminating in an EPA fuel economy equivalent of 35.5 miles per gallon (mpg) and a NHTSA fuel economy (CAFE) standard of 34.1 mpg, in model year 2016—each standard an industry fleetwide average. Various factors explain the 35.5/34.1 difference.

June 3: EPA finalizes "tailoring rule." 75 Fed. Reg. 31,514. This rule is to relieve the overwhelming permitting burdens that would, in the absence of the rule, fall on PSD and Title V permitting authorities on January 2, 2011, when EPA's light-duty vehicle rule for GHGs (see immediately above) takes effect. The tailoring rule will begin, on January 2, 2011, with GHG emissions thresholds for PSD new source review and Title V that are much higher than those in the CAA (EPA hoping to phase in the statute's low statutory thresholds after many years). Indeed, the thresholds in the final tailoring rule are higher than those in the proposed rule. For example, beginning January 2, 2011, PSD requirements will apply to projects that increase net GHG emissions by at least 75,000 tons per year CO_2 equivalent, but only if the project also significantly increases emissions of at least one non-GHG pollutant. And no source emitting less than 50,000 tons per year CO_2 equivalent will be subject to PSD new source review or Title V permitting before April 30, 2016.

August 13: EPA denies petitions to reconsider its endangerment finding for GHGs from new motor vehicles. 75 Fed. Reg. 49,556. After reviewing the 10 petitions, the agency concluded that its December 15, 2009, endangerment finding (see above) remains well-supported. Several petitions argued that emails disclosed in late 2009, many from the Climate Research Center at the University of East Anglia, in England, suggested bias among climate-change scientists, warranting a new look at the evidence for climate change.

September 2: EPA proposes PSD new source review "SIP call," and proposes FIP for states unable to timely submit corrective revisions. 75 Fed. Reg. 53,892, 53,883, respectively. EPA proposes a rule finding that EPA-approved PSD new source review programs in 13 state implementation plans (SIPs) are substantially inadequate "because they do not appear to apply PSD requirements to GHG-emitting sources." For each of these states, the same proposed rule requires it (through a SIP call) to correct its SIP. In a separate proposed rule, EPA sets out a proposed FIP for any state unable to submit, by EPA's deadline, its own SIP revision.[20] (See "December 15, 2010" below for final rule on SIP call, and "December 30, 2010" below for FIP for seven states.)

November 10: EPA issues "PSD and Title V Permitting Guidance for Greenhouse Gases." Notice of availability and solicitation of comments at 75 Fed. Reg. 70,254 (November 17, 2010); full text at http://epa.gov/regulations/guidance/byoffice-oar.html. EPA issued this guidance to assist permit writers and permit applicants in addressing the Clean Air Act's PSD and Title V permitting requirements for GHGs, which begin to apply on January 2, 2011, to certain new major stationary sources and major modifications of stationary sources (see "June 3, 2010" above: EPA finalizes "tailoring rule"). Particularly important is the guidance's discussion of the process EPA recommends for determining BACT for GHGs from such sources. (As of January 2, 2011, Clean Air Act Section 165(a)(4)[21] will require installation of such technology on certain new major stationary sources and major modifications of stationary sources proposed for PSD areas of the country.) A modified version of this guidance was issued under the date "March 2011."

December 13: EPA finalizes PSD new source review "SIP call." 75 Fed. Reg. 77,698. This final rule asserts a finding that the EPA-approved SIPs of 13 states are substantially inadequate to meet CAA requirements because they do not apply PSD requirements in their SIPs to GHGemitting sources. Owing to this finding, the rule issues a SIP call for each of the 13 states to revise its SIP as necessary to correct such inadequacies,[22] with deadlines ranging from December 22, 2010, to December 1, 2011. Note: if the state fails to correct its SIP by the deadline, the CAA requires EPA to promulgate a FIP for the state. (See "September 2, 2010" above for proposed rule.)

December 21: EPA enters into settlements agreeing to issue new source performance standards for GHG emissions from "electric generating units" (power plants) and petroleum refineries. Available at http://www.epa.gov/airquality/pdfs/boilerghgsettlement.pdf (power plants) and http://www.epa.gov/airquality/pdfs/refineryghgsettlement.pdf

(petroleum refineries). The two settled lawsuits were petitions for review of EPA amendments to its existing new source performance standards (NSPSs) for, respectively, electric generating units[23] and petroleum refineries.[24] On each occasion, petitioners objected, EPA had declined to introduce NSPSs for GHG emissions. In the settlements, EPA agrees to (a) propose by July 26, 2011, NSPSs for GHG emissions from new/modified electric generating units and guidelines for existing electric generating units, then promulgate final NSPSs and guidelines by May 26, 2012, and (b) propose by December 10, 2011, NSPSs for GHG emissions from new/modified petroleum refineries and guidelines for existing petroleum refineries, then promulgate final NSPSs and guidelines by November 10, 2012. EPA may withdraw its approval of either settlement within 30 days after the public comment period. Note: EPA has not yet proposed either the NSPSs for electric generating units or for petroleum refineries, and thus has missed both proposal deadlines in the settlements.

December 30: EPA finalizes rule to narrow previous approval of state Title V permitting programs that apply to GHG-emitting stationary sources. 75 Fed. Reg. 82,254. This rule is a companion to that below. It narrows EPA's previous approval of state Title V operating permit programs so that only stationary sources that exceed the GHG thresholds established in the "tailoring rule" (see "June 3, 2010" above) are covered as major sources by the federally approved Title V programs in the affected states. By thus raising the GHG emissions thresholds that apply Title V permitting to major sources in the affected states, this rule aims to reduce the number of stationary sources that will be required to have Title V permits, and thereby reduce Title V permitting burdens for state permitting agencies and sources in the affected states.

December 30: EPA finalizes rule to narrow previous approval of SIP PSD programs that apply to GHG-emitting stationary sources. 75 Fed. Reg. 82,536. This rule is a companion to that above. It narrows EPA's previous approval of SIP PSD programs that apply to GHG-emitting stationary sources, by withdrawing approval of those programs to the extent they apply PSD to GHG-emitting sources below the thresholds in the "tailoring rule" (see "June 3, 2010" above). By thus raising the thresholds in 24 states above the statutory threshold, this rule aims to reduce the number of new stationary sources, or major modifications of existing sources, that will be required to have PSD permits, and thereby reduce PSD permitting burdens for state permitting agencies and sources in the affected states.

December 30: EPA establishes GHG PSD federal implementation plans for seven states. 75 Fed. Reg. 82,246. Following up on the December

13, 2010, SIP call (see above), this EPA rule finalizes a FIP to apply in each of the seven states that did not submit by the December 22, 2010, EPA deadline a revised SIP to apply their EPA-approved PSD program to GHG emissions. These states are: Arizona (most of state), Arkansas, Florida, Idaho, Kansas, Oregon, and Wyoming. Other states have later deadlines. The FIP ensures that a permitting authority—that is, EPA—is available in these states as of January 2, 2011, the date when the CAA's PSD requirements begin applying to GHG-emitting sources (see "April 2, 2010" for how January 2, 2011, was chosen). Without a permitting authority to issue CAA-required permits, proposed major stationary sources and major modifications of existing sources could not begin construction if they would emit GHGs in quantities above the tailoring rule thresholds. (See "September 2, 2010" for proposed FIP.)

December 30: EPA converts previous full approval of Texas's PSD program to partial approval, and promulgates FIP applying PSD to large GHG-emitting stationary sources. 75 Fed. Reg. 82,430. In this rule, EPA finds that it erred when it fully approved Texas's PSD program in 1992, since the program did not address its application to all pollutants newly subject to regulation, including GHGs. As a result, this rule changes EPA's 1992 approval from full to partial.[25] This change requires EPA under the CAA to issue a FIP,[26] which the rulemaking also does. Under the FIP, EPA will become the permitting authority for proposed GHG-emitting stationary sources in Texas in accordance with the emissions thresholds in the tailoring rule. Without a permitting authority to issue CAA-required permits, proposed major stationary sources and major modifications of existing sources in Texas could not begin construction if they would emit GHGs in quantities above the tailoring rule thresholds.

2011

July 20: EPA finalizes rule deferring application of PSD and Title V permitting requirements to CO2 emissions from bioenergy and other biogenic stationary sources. 76 Fed. Reg. 43,490. Such CO2 emissions are generated by combustion or decomposition of biologically based material—as at solid waste landfills, manure management operations, and electric utilities burning biomass fuels. The deferral, to allow EPA more time to examine how to account for such emissions, is for three years. During this period, biogenic emissions are not required to be counted for determining whether a source is subject to PSD and Title V permitting. The deferral applies only to CO2 emissions and does not affect non-GHG pollutants or other GHGs emitted from the combustion of biomass fuel. EPA is taking this action as part of granting the petition for reconsideration filed by the National Alliance of Forest Owners on August 3, 2010, related to the tailoring rule (see "June 3, 2010" above).

September 15: EPA and NHTSA jointly finalize rules setting GHG emission standards and fuel economy standards for 2014 and later model year medium- and heavy-duty vehicles. 76 Fed. Reg. 57,106. These rules, weighing in at 958 pages (including preamble), respond to a presidential memorandum of May 21, 2010.[27] EPA's emission standards and NHTSA's fuel economy standards apply to three categories of heavy-duty vehicles: combination tractors, heavy-duty pickup trucks and vans, and vocational vehicles. The rules include separate standards for the engines that power combination tractors and vocational vehicles. Certain rules are exclusive to EPA, such as EPA's hydrofluorocarbon standards to control leakage from air conditioning systems in combination tractors, and pickup trucks and vans. EPA's emission standards will begin with model year 2014. NHTSA's fuel economy standards are voluntary in model years 2014 and 2015, becoming mandatory for most vehicle categories in model year 2016.

December 1: EPA and NHTSA jointly propose rules setting GHG emission standards and fuel economy standards for 2017-2025 model year light-duty vehicles. 76 Fed. Reg. 74,854. These rules, 893 pages long (including preamble), respond to a presidential memorandum of May 21, 2010.[28] The standards will apply to passenger cars, light-duty trucks, and medium-duty passenger vehicles and will build on the model year 2012-2016 light-duty-vehicle standards (see "May 7, 2010" above). California and 13 auto manufacturers have provided letters of support for this new phase. EPA's emission standards will be more stringent each year from 2017 to 2025, achieving, as an industry fleetwide average, the equivalent of 54.5 miles per gallon (mpg) in model year 2025. NHTSA's CAFE standards will increase annually and require, as an industry fleetwide average, 49.6 mpg in model year 2025. Various factors explain the 54.5/49.6 difference.

See entry for May 7, 2010, for statutory and historical background.

2012

March 8: EPA proposes under the tailoring rule to leave the GHG major source thresholds unchanged. 77 Fed. Reg. 14226. Under step 1 of the tailoring rule (see "June 3, 2010"), which began on January 2, 2011, stationary sources above the tailoring-rule GHG threshold that are required to obtain a PSD or Title V permit anyway due to emissions of other pollutants must address their GHG emissions in the permit. Under step 2, effective July 2, 2011, sources with GHG emissions above the tailoring-rule threshold also are required to obtain a PSD or Title V permit, even if they would not fall under these programs based on emissions of other pollutants. In this proposal, EPA finds that the capabilities of state permitting authorities have not improved enough for additional GHG sources to be brought under PSD or Title V through step 3. Therefore, EPA proposes to leave the GHG major source thresholds unchanged at this time. EPA also proposes two "streamlining" approaches to improve the administration of the GHG PSD and Title V permitting programs.

April 13: EPA proposes new source performance standards for CO2 emissions from new fossil-fuel-fired electric generating units. 77 Fed. Reg. 22,392. The proposed rule, pursuant to CAA Section 111, would require new fossil-fuel-fired electric generating units (power plants) of greater than 25 megawatt capacity to meet an output-based standard of 1,000 pounds of CO2 per megawatt-hour—a standard based, according to EPA, on the performance of widely used natural gas combined cycle technology. EPA had committed to issuing this proposed rule by July 26, 2011, in a settlement of litigation. See "December 21, 2010."

APPENDIX. TABLE OF ACRONYMS

BACT	Best available control technology.
	This is the pollution control standard in PSD areas.
	BACT is defined in Clean Air Act Section 169(3).
CAA	Clean Air Act. Codified at 42 U.S.C. §§7401-7671q.
CAFE	Corporate average fuel economy.
	The text uses "CAFÉ standards" synonymously with "fuel economy standards."
CEQ	Council on Environmental Quality, the agency charged with monitoring executive branch implementation of the National Environmental Policy Act. 42 U.S.C. §§4342, 4344.
EPA	Environmental Protection Agency

FIP	Federal implementation plan. Clean Air Act Section 110(c) requires EPA to issue a FIP for a state, setting out emission limits for stationary sources in the state, whenever the state fails to submit a required SIP or a state-submitted SIP does not meet minimum criteria.
GHG	Greenhouse gas
NEPA	National Environmental Policy Act. 42 U.S.C. §4321 et seq. NEPA requires federal agencies to prepare environmental impact statements for proposed "major federal actions significantly affecting the quality of the human environment." *Id.* at §4332(2)(C).
NHTSA	National Highway Traffic Safety Administration
NSPS	New source performance standards. These apply to any stationary source of emissions the construction or modification of which is begun after the NSPS is proposed. Defined in Clean Air Act Section 111(a)(1).
PSD	Prevention of significant deterioration. Under the Clean Air Act, PSD areas are regions where the ambient concentration of a pollutant is below (cleaner than) the National Ambient Air Quality Standard for that pollutant, triggering the act's regime for "preventing significant deterioration" of that air quality. See 42 U.S.C. §§7470-7492.
SEC	Securities and Exchange Commission
SIP	State implementation plan. Clean Air Act Section 110(a)(1) requires each state to submit a SIP to EPA to achieve each National Ambient Air Quality Standard within that state. The state has discretion in imposing emission limits on stationary sources within the state as long as the National Ambient Air Quality Standard is achieved.

End Notes

[1] 549 U.S. 497 (2007).
[2] CAA §202(a); 42 U.S.C. §7521(a).
[3] *Id.*
[4] CAA §165(a)(4); 42 U.S.C. §7475(a)(4).
[5] CAA §§501-507; 42 U.S.C. §§7661-7661f.
[6] *See* CRS Report RL32764, *Climate Change Litigation: A Survey*, by Robert Meltz; CRS Report R40984, *Legal Consequences of EPA's Endangerment Finding for New Motor Vehicle*

Greenhouse Gas Emissions, by Robert Meltz; CRS Report RS22665, *The Supreme Court's Climate Change Decision: Massachusetts v. EPA*, by Robert Meltz; CRS Report R40585, *Climate Change: Potential Regulation of Stationary Greenhouse Gas Sources Under the Clean Air Act*, by Larry Parker and James E. McCarthy; CRS Report R40506, *Cars, Trucks, and Climate: EPA Regulation of Greenhouse Gases from Mobile Sources*, by James E. McCarthy and Brent D. Yacobucci; and CRS Report R40166, *Automobile and Light Truck Fuel Economy: The CAFE Standards*, by Brent D. Yacobucci.

[7] Robin Bravender, *With Hill hopes dashed, advocates circle wagons at EPA*, Greenwire (August 25, 2010). Characterizing EPA's efforts as "the only game in town" for dealing with climate change is not *strictly* accurate, unless "town" is taken literally to mean Washington, DC. Outside Washington, DC, entities unsatisfied with the pace of congressional and EPA action vis-à-vis climate change have looked to international forums, treaty negotiations, state and regional efforts, and lawsuits seeking to establish GHG emissions as a common law nuisance. Regarding the last item, see CRS Report R41496, *Common-Law Climate Change Litigation After American Electric Power v. Connecticut*, by Robert Meltz.

[8] These lawsuits challenge primarily four GHG-related actions by EPA. The four actions, and the dates under which they are described in this report, are (1) the "endangerment finding" (December 15, 2009); (2) the "tailpipe rule" (May 7, 2010); (3) the "timing rule" (April 2, 2010); and (4) the "tailoring rule" (June 3, 2010). See Gregory E. Wannier, *EPA's Impending Greenhouse Gas Regulations: Digging Through the Morass of Litigation*, available at http://www.law.columbia.edu/centers/climatechange/publications/ working papers.

[9] CAA §209(a); 42 U.S.C. §7543(a).

[10] CAA §209(b); 42 U.S.C. §7543(b).

[11] *Id.*

[12] CAA §177; 42 U.S.C. §7507.

[13] CAA §209(b)(1)(B); 42 U.S.C. §7543 (b)(1)(B).

[14] CAA §165(a)(4); 42 U.S.C. §7475(a)(4).

[15] P.L. 110-161, Div. F, tit. II; 121 Stat. 1844, 2128.

[16] 42 U.S.C. §7521(a).

[17] 42 U.S.C. §4321 et seq.

[18] 49 U.S.C. §32902(b)(2)(A).

[19] *See* http://www.arb.ca.gov/regact/2010/ghgpv10/ghgpv10.htm.

[20] EPA is required to promulgate a FIP when a state fails to make a required SIP revision. CAA §110(c)(1), 42 U.S.C. §7410(c)(1).

[21] 42 U.S.C. §7475(a)(4).

[22] As required by CAA §110(k)(5), 42 U.S.C. §7410(k)(5).

[23] 71 Fed. Reg. 9,866 (2006).

[24] 73 Fed. Reg. 35,838 (2008).

[25] See CAA §110(k)(6), 42 U.S.C. §7410(k)(6).

[26] See CAA §110(c)(1), 42 U.S.C. §7410(c)(1).

[27] See note 22 *supra*.

[28] See note 22 *supra*.

In: Climate Change: Legal Issues and Contexts ISBN: 978-1-62257-847-4
Editors: B. Saunders and R. A. Diaz © 2013 Nova Science Publishers, Inc.

Chapter 4

CLIMATE CHANGE LITIGATION: A SURVEY[*]

Robert Meltz

SUMMARY

The scientific, economic, and political questions surrounding climate change have long been with us. This report focuses instead on a relative newcomer: the legal debate. Though the first court decision related to climate change appeared 19 years ago, such litigation has proliferated in just the past six. Representatives of some suing organizations and states acknowledge that a prime cause for this litigation surge was inaction by Congress and the executive branch during the George W. Bush Administration with regard to mandatory constraints on greenhouse gas (GHG) emissions.

The court cases, decided and pending, arise in eight contexts. The first is the Clean Air Act (CAA). In *Massachusetts v. EPA*, the Supreme Court held that as to mobile sources of emissions (cars, trucks), EPA has authority under the act to regulate greenhouse gas (GHG) emissions. This decision puts pressure on EPA to move forward as well with regulation of GHGs from stationary sources (power plants, factories).

Second, litigation under wildlife statutes, particularly the Endangered Species Act, raises the possibility that the impacts of climate change on wildlife may constrain private activities that emit GHGs.

---

[*] This is an edited, reformatted and augmented version of Congressional Research Service, Publication No. RL32764, dated April 15, 2009.

Third, energy statutes have been invoked. It has been held, for example, that under the Energy Policy and Conservation Act, the United States must monetize the benefits of reduced carbon emissions as part of setting light-truck fuel economy standards.

Fourth, various statutes requiring federal government analysis and information dissemination— the National Environmental Policy Act (NEPA), Global Change Research Act, and Freedom of Information Act—have generated climate-change litigation. NEPA suits make up the most numerous subset of this category. Courts agree that if a plaintiff can establish standing, NEPA can be used to compel agency consideration of the climate change effects of its actions.

Fifth, common law tort theories such as nuisance have been invoked, not yet successfully, to force cutbacks in GHG emissions, or payment of damages. Several cases are on appeal.

Sixth are the preemption suits. These challenge state regulation of GHG emissions from motor vehicles as preempted by the federal corporate average fuel economy standards or federal authority over foreign policy. The two rulings thus far have rejected these challenges, but are on appeal. California's suit attacking EPA's denial of its request for a waiver of federal preemption under the Clean Air Act has now been stayed, pending EPA reconsideration of the denial.

Seventh, chiefly with respect to coal-fired power plants, are suits under state utilities laws.

And eighth, one case asks whether existing general liability insurance policies cover climatechange-related liability.

Finally, the report discusses international law aspects of a nation's contributions to climate change, and offers some overview comments.

INTRODUCTION

The scientific, economic, and political questions surrounding climate change have long been with us. This report focuses instead on a relative newcomer: the legal debate. Though the first court decision related to climate change appeared 19 years ago, the quantity of such litigation has mushroomed in recent years. One observer counts 118 lawsuits and petitions for government action filed on climate change issues through the end of 2008—41 lawsuits filed in 2007 alone.[1]

Representatives of some suing organizations and states acknowledge that a prime cause for this litigation surge was inaction by Congress and the executive branch during the George W. Bush Administration with regard to mandatory constraints on greenhouse gas (GHG) emissions, and their perception that litigation might help to prompt such action.

The principal court cases, decided and pending, arise in eight contexts—a number that continues to grow. First and most important is the Clean Air Act (CAA). In April, 2007, the Supreme Court held in *Massachusetts v. EPA* that EPA has authority under the CAA to regulate greenhouse gas emissions from new motor vehicles.[2]

The second context for climate change litigation is the federal wildlife statutes, raising the issue of whether statutes like the Endangered Species Act can be used to limit GHG emissions based on their contribution to climate-climate-related alterations of wildlife habitat.

Third is the federal energy statutes, such as the Energy Policy and Conservation Act and Outer Continental Shelf Lands Act, which also raise questions as to whether climate change impacts must be considered in their spheres. The fourth context for litigation is federal information statutes such as the National Environmental Policy Act, exploring the extent to which they can be used to compel government analysis of and dissemination of information about climate change.

Fifth is common law tort theories such as nuisance and whether they be used successfully by state and private plaintiffs to force cutbacks in GHG emissions, or payment of damages? Sixth is federal preemption of state regulation of GHG emissions.

This category breaks down into efforts by states and environmentalists to reverse EPA's refusal to waive Clean Air Act preemption, and auto industry efforts to impose preemption under non-Clean Air Act theories, such as the "CAFE standards" under the Energy Policy and Conservation Act.

Seventh, chiefly with respect to coal-fired power plants, is state utilities laws. And eighth is whether general liability insurance policies cover harms and liabilities caused by climate change. Sections I through VIII of this report address these eight areas of litigation in turn.[3] Most known cases, decided and pending, are mentioned—omitted cases are those that raise climate change issues in only the most marginal way or only implicitly,[4] and some state cases. Looking beyond the domestic lawsuits, Section IX surveys international law arguments that might be used to induce GHG emission reductions from the United States and other countries that are major GHG emitters, and the few international law claims filed against the United States to date. Finally, Section X offers overall comments.

I. CLEAN AIR ACT

Stationary Sources of GHG Emissions

The First EPA General Counsel Memorandum

Aware that prospects for Senate approval of the Kyoto Protocol were dubious,[5] some Members of Congress became concerned in the late 1990s that the Clinton Administration EPA might seek to regulate GHG emissions in the absence of approval, under either of two claimed authorities. One authority would derive from an argument that even prior to ratification, the Protocol provided some sort of legal basis for emissions restrictions, perhaps citing past treaties signed by the United States that were provisionally implemented prior to going into effect.[6] This possibility provoked a series of enactments barring EPA's use of appropriated funds to implement the Kyoto Protocol in the absence of approval and ratification.[7]

The rest of this section deals with the second claim of possible authority: regulating GHG emissions independently of the Protocol, under the CAA. This authority has now been confirmed by the Supreme Court, at least as to mobile sources; nonetheless, this report retains from earlier versions the historical evolution of the issue.

In 1998, an EPA General Counsel memorandum[8] concluded that CO_2 satisfies the CAA definition of "air pollutant," but that this conclusion is only the first step. Before EPA can regulate CO_2 emissions, the memorandum went on, it must further conclude that CO_2 meets criteria in other CAA provisions requiring the agency to determine that the substance poses harm to public health, welfare, or the environment. This next step EPA declined to take. At a House hearing in 1999,[9] a panel of legal experts argued the question of EPA's authority to regulate CO_2 under the CAA. A new EPA General Counsel endorsed his predecessor's analysis in the 1998 memorandum, but just as his predecessor, stressed that the EPA's legal analysis was "largely theoretical" since "EPA currently has no plans to regulate carbon dioxide...."[10] This hands-off position was prompted in part by strong congressional opposition based on uncertainties as to the economic impact of regulating a pollutant as widespread as CO_2. In addition, some in Congress argued that CAA implementation of a CO_2 standard was barred by the aforementioned enactments (appropriation riders) prohibiting implementation of the Kyoto Protocol.[11]

The EPA General Counsel opinion that "air pollutant" includes GHGs held sway until 2003, when that office reversed itself in the context of a petition asking the agency to regulate GHG emissions from mobile sources. This story picks up below ("Mobile Sources of GHG Emissions").

Suits Seeking General CAA Rulemaking by EPA

The earliest lawsuit in this category, now dismissed, sought to have EPA promulgate national ambient air quality standards for CO_2. In *Massachusetts v. Whitman*, filed in 2003, three Northeast states (MA, CT, ME) sought to force EPA to list CO_2 as a "criteria pollutant" under the CAA.[12] They argued that on various occasions, EPA had indicated its belief that CO_2 emissions contribute to climate change. These EPA statements constituted, in the words of CAA section 108,[13] a "judgment [that GHG emissions] cause or contribute to air pollution which may reasonably be anticipated to endanger public health or welfare" and "result[] from numerous or diverse mobile or stationary sources." These prerequisites being satisfied, the suit argued, section 108 requires EPA to add CO_2 to its list of "criteria pollutants," then proceed under section 109[14] to develop national ambient air quality standards for CO_2. On September 3, 2003, a few days after EPA's denial of a petition asking the agency to regulate GHG emissions from motor vehicles, the plaintiff states voluntarily dismissed this suit, reportedly so as to transfer their energies to a suit challenging the petition denial (leading to the Supreme Court's *Massachusetts v. EPA* decision).

The remaining suits in this category are all active today. Each one seeks EPA regulation of GHG emissions through new source performance standards (NSPSs) under the CAA.[15] In most of these cases, regulation of GHGs is not the primary issue. Nonetheless, it should be noted that they will be litigated in the shadow of the Supreme Court ruling in *Massachusetts v. EPA* holding that EPA has authority under the CAA to regulate GHGs from mobile sources. The burning question is how that ruling will affect EPA regulation of *stationary-source* GHG emissions.

The first NSPS suit was *New York v. EPA*, an effort to compel EPA to issue a NSPS for CO_2 from steam generating units. *New York* began with an EPA proposal to revise its NSPSs for electric utility and other steam-generating units. Some commenters on the proposed rule argued that EPA must, in addition to the revisions proposed, set NSPSs for GHGs emitted from steam generating units. The commenters pointed to CAA section 111's command that EPA promulgate NSPSs to address emissions from new stationary sources that "cause[], or contribute[] significantly to, air pollution which may reasonably be anticipated to endanger public health or welfare." In promulgating its final rule in February, 2006,[16] however, EPA rejected this demand, saying (it being pre-*Massachusetts v. EPA*) that the agency lacked authority to set NSPSs for GHGs. Review of the final rule was sought in the D.C. Circuit.[17] In 2006, the court severed the portion of the case dealing with regulation of GHGs, titling it *New York v. EPA*.[18] In 2007, a few months after the Supreme Court decision in *Massachusetts v. EPA*, this severed case was remanded to EPA for further proceedings in light of that decision (this section *infra*).

Two other NSPS suits concern oil and natural gas—the production side in one suit; the refining side in the other. As to production, plaintiffs in *WildEarth Guardians v. Johnson*, No. 1:09 CV 00089 (D.D.C. filed January 14, 2009), invoke the CAA citizen suit provision to force EPA performance of alleged nondiscretionary duties under both the NSPS and National Emission Standards for Hazardous Air Pollutants provisions of the act.[19] The complaint notes CO_2, and particularly methane, as GHG pollutants of concern. Since neither of these pollutants is listed as a hazardous air pollutant, it may be surmised that the sole portion of the lawsuit pertinent to climate change is the NSPS claim. That claim alleges EPA's failure to review and revise the NSPS for the Crude Oil and Natural Gas Production category since 1985, despite the CAA requirement that NSPSs be reviewed, and if appropriate revised, every eight years.[20]

As for oil and gas refining, New York and 11 other states (CA, CT, DE, ME, MA, NH, NM, OR, RI, VT, and WA) filed a petition for review in 2008 challenging EPA's revisions that year of its existing NSPS for petroleum refineries.[21] Petitioners' argument in *New York v. EPA*, No. 08-1279 (D.C. Cir. filed August 25, 2008), is that EPA acted arbitrarily and capriciously in failing to determine whether GHGs from petroleum refineries endanger public health and welfare and by failing to promulgate NSPSs for GHG emissions in the refinery rule. Petitioners cite CAA section 111(b)'s requirement that EPA make an endangerment determination, establish NSPS for a source category that contributes significantly to such endangerment, and revise the NSPS at least every eight years. This lawsuit has been consolidated with several others, including *Environmental Integrity Project v. EPA* (No. 08-1281), and is now styled *American Petroleum Institute v. EPA* (No. 08-1277). On December 15, 2008, the D.C. Circuit granted a motion to hold the case in abeyance while EPA considers various petitions for administrative reconsideration of the rule.[22]

The most recent NSPS suit, *Environmental Integrity Project v. U.S. EPA*, No. 1:09-CV 00218 (D.D.C. filed February 4, 2009), seeks to compel EPA to perform its nondiscretionary duty to review, and if necessary revise, its NSPS for nitric acid plants, including for nitrous oxide (N_2O), a powerful GHG.. The standard, plaintiffs allege, has not been reviewed since 1984, notwithstanding the CAA requirement that NSPSs be reviewed and if appropriate revised every eight years.[23]

Suits and Administrative Petitions Enforcing the CAA Against Specific Stationary Sources

Rather than seeking general rulemaking as above, these legal efforts seek to impose CO_2 emission limits on specific manufacturing, power, and heating plants.

In *Northwest Environmental Defense Center v. Owens Corning Corp.*, environmental groups invoke the CAA citizen suit provision to enforce the act's "new source review" requirement as to GHG emissions.[24] They contend that Owens Corning is constructing a manufacturing plant in Oregon with the potential to emit more than 250 tons per year of harmful gases, without having obtained the required CAA permit.[25] The principal such gas is HCFC-142b, which plaintiffs contend is a potent GHG. In a preliminary ruling, the court held that plaintiffs have standing, notwithstanding that the climate change impacts of the plant's GHG emissions would be "indirect." Anticipating the Supreme Court's rationale for granting standing in *Massachusetts v. EPA*, the court found that standing was not precluded by the fact that the injury to plaintiffs would be shared with many others, nor because the relief sought would not lead to a *complete* elimination of climate change impacts.

At least three proceedings have involved Sierra Club appeals to EPA's Environmental Appeals Board of recently issued "new source review" permits in Prevention of Significant Deterioration (PSD) areas.[26] In each case, the issue has been whether a permit issued for construction of a new "major emitting facility" or major modification in a PSD area must require the use of "best available control technology" (BACT) for CO_2 emissions from that source.[27] In two places, the CAA requires such new facilities or major modifications to install BACT for "each pollutant subject to regulation under this Act."[28] As indicated below, Sierra Club has used two arguments in support of its position that GHGs are "pollutant[s]subject to regulation" by the statute.

In the first proceeding, *In re Christian County Generation, LLC,* PSD Appeal No. 07-01, 13 E.A.D. ____ (January 28, 2008), Sierra Club objected to the issuance of a PSD permit by a state agency for construction of a coal-fired electric power plant. The Board denied the petition because Sierra Club had not raised its argument during the public comment period on the draft permit. Sierra Club argued that the comment period closed before the Supreme Court decision in *Massachusetts v. EPA* holding that GHGs are "air pollutants" under the CAA, but the EAB found that the arguments in that case were reasonably ascertainable at the time of the public comment period.

The second and most publicized case, *In re Deseret Power Electric Cooperative*, PSD Appeal No. 07-03, 14 E.A.D. _____ (November 13, 2008), involved a PSD permit issued by EPA Region 8 to allow the construction of a new waste-coal-fired utility generating unit at an existing power plant located near Bonanza, Utah. The Board rejected Sierra Club's contention that "subject to regulation" has a plain meaning compelling Region 8 to impose a CO_2 BACT limit in the PSD permit. Sierra Club had pointed to EPA's 1993 amendment of its regulations requiring monitoring and reporting of CO_2 emissions,[29] as directed by section 821 of the 1990 CAA Amendments. At the same time, the Board rejected the Region's argument that it was limited by an historical agency interpretation to read "subject to regulation" as meaning "subject to a statutory or regulatory provision that requires actual control of emissions of that pollutant." Since EPA has yet to issue a CAA regulation requiring actual control of CO_2 emissions, Region 8 argued, BACT for CO_2 is not required. Hence, the Board remanded the permit to the Region for it to reconsider whether to impose a CO_2 BACT limit. The Board recognized, however, that given the significance of the issue, it would be best if the Agency, rather than one of its regional offices, defined "subject to regulation under this Act."

This the EPA Administrator did a month later. In a memorandum issued December 18, 2008, he declared EPA's "definitive interpretation" of its regulation defining which pollutants trigger new source review in PSD areas.[30] Reprising Region 8's argument in *Deseret Power*, the Administrator said that this regulation "exclude[s] pollutants for which EPA regulations only require monitoring or reporting but ... include[s] each pollutant subject to either a provision in the Clean Air Act or regulation adopted by EPA under the Clean Air Act that requires actual control of emissions of that pollutant." To reiterate, this position excludes CO_2 until EPA promulgates a regulation covering CO_2 emissions. Sierra Club petitioned for review of the memorandum, which was granted by the newly arrived Obama Administration on February 17, 2009.

In the third EAB appeal, *In re Northern Michigan University (Ripley Heating Plant)*, PSD Appeal No. 08-02, 14 E.A.D. _____ (February 18, 2009), Sierra Club challenges a permit issued by the Michigan Department of Environmental Quality to the university, allowing it to construct a new circulating fluidized bed boiler at the heating plant.[31] Sierra Club's argument was the same as in *Deseret*, and the Board ruled identically—that is, it directed the Michigan agency, guided by *Deseret*, to consider whether "pollutant subject to regulation" requires application of a BACT limit to CO_2 emissions. The Board's decision makes no mention of the Administrator's December, 2008 memo or EPA's grant of review thereof in 2009.

Mobile Sources of GHG Emissions

The Section 202 Petition Denial and the Second EPA General Counsel Memorandum

In 1999, 19 organizations petitioned EPA to regulate emissions of GHGs (CO_2, methane, nitrous oxide, and hydrofluorocarbons) from new motor vehicles. The rulemaking petition cited the agency's alleged mandatory duty to do so under CAA section 202(a)(1).[32] That section directs the EPA Administrator to prescribe emission standards for "any air pollutant" from new motor vehicles "which, in his judgment cause[s], or contribute[s] to air pollution which may reasonably be anticipated to endanger public health or welfare."

In 2003, EPA denied the section 202 petition.[33] Much of the agency's rationale followed a new General Counsel memorandum, issued the same day.[34] Contrary to its Clinton Administration precursor, this new OGC memorandum concluded that the CAA does *not* grant EPA authority to regulate CO_2 and other GHG emissions for their climate change impacts.

Massachusetts v. EPA: The Challenge to EPA's Petition Denial

EPA's denial of the section 202 petition prompted a suit in the D.C. Circuit by twelve states (CA, CT, IL, MA, ME, NJ, NM, NY, OR, RI, VT, WA) and others. Opposing the challenge, besides EPA, were ten state intervenors (AK, ID, KS, MI, ND, NE, OH, SD, TX, UT), plus several automobile- and truck-related trade groups. In 2005, a split panel in *Massachusetts v. EPA* rejected the suit,[35] and the Supreme Court granted certiorari.

In *Massachusetts v. EPA*, the Supreme Court ruled 5-4 for petitioners on all three issues in the case.[36] First, Massachusetts, the majority held, had standing to bring the claim. Second, EPA has authority to regulate motor vehicle GHGs under section 202, since "air pollutant" includes GHG emissions. And third, the phrase "in his judgment" in section 202 does not allow EPA to exercise discretion against regulating based on policy considerations. The ruling in favor of petitioners was forecast early in the majority opinion by its opening sentences: "A well-documented rise in global temperatures has coincided with a significant increase in the concentration of carbon dioxide in the atmosphere. Respected scientists believe the two trends are related."[37] (Nor did the dissenters dispute this.)

Most of the decision is devoted to the first issue, standing. The Court found that petitioners had two factors in their favor. First, the CAA specifically authorizes challenges to agency action unlawfully withheld, such as here.[38] A litigant to whom Congress has accorded such a procedural right, said the Court, "can assert that right without meeting all the normal standards for redressability and immediacy"[39]—two prerequisites of the standing test. Second, the Court found it "of considerable relevance"[40] that the petitioner injury on which it focused—Massachusetts's loss of shore land from global-warming-induced sea level rise—was that of a sovereign state. States are "not normal litigants for the purposes of invoking federal jurisdiction,"[41] said the Court, noting their quasi-sovereign duty to preserve their territory.

Having described petitioners' favored position in establishing standing, it was surprising that the Court then undertook a traditional standing analysis. As to the first prong of the standing test— whether plaintiff has demonstrated actual or imminent "injury in fact" of a concrete and particularized nature— the Court homed in on Massachusetts's status as *owner* of much of the shore land being lost to sea level rise. That this injury may be widely shared with other coastal states does not disqualify this injury, said the Court; it is nonetheless concrete.

The second prong of the standing test is causation, requiring that the injury of which the plaintiff complains is fairly traceable to the defendant. EPA did not dispute the existence of a causal relationship between GHG emissions and climate change. It did argue, however, that any reduction in GHG emissions achieved through the current litigation would be too small a portion of worldwide GHG emissions to make a cognizable difference in climate change. In a ruling that may be of benefit to environmental plaintiffs in many contexts, the Court held that even an agency's refusal to take a "small incremental step"[42] that would result in only a modest reduction in worldwide GHG emissions, is enough for standing purposes.

The third and final prong of the standing test is redressability, demanding that the remedy sought by the plaintiff is one that is likely to redress his/her injury. Here, the remedy sought was EPA regulation of GHG emissions from new motor vehicles. The Court found that this remedy satisfied redressability because while it would not by itself reverse climate change, it would nonetheless slow or reduce such change. Nor, given the "enormity"[43] of the potential effects of climate change, was it relevant to the Court that the full effectiveness of the remedy would be delayed until existing cars and trucks on the road were largely replaced by new ones.

In contrast with the Court's lengthy discourse on standing, its handling of the CAA issues in the case is quite brief. On the authority question, the Court said that the CAA's broad definition of "air pollutant" simply cannot be squared with EPA's view that GHGs are not included. The Court rejected EPA's argument that federal laws enacted following enactment of this definition—laws emphasizing interagency collaboration and research—suggest that Congress meant to curtail EPA's power to use mandatory regulations in addressing air pollutants. Nor was the Court impressed with EPA's argument that "air pollutant" in the CAA could not include vehicle GHG emissions because EPA standards for such emissions could be satisfied only by improving fuel economy, a goal assigned to the Department of Transportation under a different statute (the Energy Policy and Conservation Act[44]).

Finally, on the discretion issue, the majority concluded that "in his judgment" refers only to whether an air pollutant "may reasonably be anticipated to endanger public health or welfare." Thus, said the Court, EPA can avoid taking further action in response to the section 202 petition "only if it determines that greenhouse gases do not contribute to climate change or if it provides some reasonable explanation as to why it cannot or will not exercise its discretion." It rejected EPA's stated policy reasons for refusing to regulate GHG emissions, such as its claim that voluntary executive branch programs already provide an effective response to climate change and that unilateral EPA regulation of vehicle GHG emissions could weaken U.S. efforts to persuade developing countries to reduce the GHG intensity of their economies. Such reasons "have nothing to do with whether greenhouse gas emissions contribute to climate change."[45] In short, said the Court, the only question is whether sufficient information exists to make an endangerment finding under section 202.

Accordingly, the Supreme Court reversed the D.C. Circuit opinion and remanded the case to that court for further proceedings.[46] On September 14, 2007, the D.C. Circuit vacated EPA's denial of the section 202 petition and remanded the matter to the agency. (Further developments are described in the following "Aftermath" section.)

A four-justice dissent by Chief Justice Roberts in *Massachusetts v. EPA* disputed the majority's holding of standing. A dissent by Justice Scalia for the same four justices argued that agency policy preferences may appropriately be considered as part of EPA's decision *whether* to issue a "judgment," conceding that the judgment, *if made*, must be limited to whether vehicle GHG emissions cause endangerment. Justice Scalia also disputed the majority's holding that "air pollutant" in section 202 includes GHGs.

Aftermath of Supreme Court Decision

The Court's decision left EPA with three options: make a finding that motor vehicle GHG emissions may "endanger public health or welfare" and issue emissions standards; make a finding that such emissions do not satisfy that prerequisite; or decide that climate change science is so uncertain as to preclude making a finding either way (or cite some other "reasonable explanation" why it will not exercise its discretion either way).[47] As to the state of climate change science, the Court's focus on the policy reasons EPA gave as part of exercising its "judgment" obscures the fact that the agency's rejection of the petition stemmed in part from expressions of scientific uncertainty in a 2001 National Research Council report on the science of climate change. Whether scientific reports since the petition rejection in 2003 have foreclosed the scientific-uncertainty rationale is beyond the scope of this report.[48]

The EPA Administrator did say after the decision that although it bars EPA use of policy considerations as a basis for *denying* the petition, it left open whether the agency can invoke them when actually *writing* the regulations, should the agency make an endangerment finding.[49] CAA section 202 does not impose any stringency or other criteria on GHG emission standards promulgated under the section. Given the wide latitude EPA has in setting section 202 standards for GHGs, the possibility exists that EPA, following an endangerment finding, could set voluntary standards, or standards pegged to the CAFE standards for fuel economy, or standards that must be complied with only after the President certifies that developing nations have put adequate GHG emission limits into effect.

In May, 2007, President Bush asked EPA to have CAA regulations limiting vehicle GHG emissions in place by the end of 2008 and to use the President's 2007 State of the Union proposal for raising the CAFE standards as a guide.[50] As late as early December, 2007, EPA was consistently stating that it intended to issue proposed regulations by the end of 2007. However, the enactment of the Energy Independence and Security Act in December, 2007,[51] with its increase in CAFE standards, led EPA to back off from any firm deadline for issuance of mobile-source GHG emission standards. In early

2008, EPA proposed instead to issue an advance notice of proposed rulemaking (ANPR) addressing the full range of *Massachusetts v. EPA*'s ramifications throughout the CAA, not just on section 202 standards. In response, the *Massachusetts v. EPA* petitioners in April, 2008 requested the D.C. Circuit to order EPA to comply with the Supreme Court's remand and the Circuit's mandate within 60 days (by choosing one of the three options noted earlier). The court denied the request in June, 2008.[52] The following month, EPA issued a lengthy ANPR that, it said, "reflects the complexity and magnitude of the question of whether and how greenhouse gases could be effectively controlled under the Clean Air Act"[53]—extending well beyond the narrow section 202 endangerment issue in the case. It warned that regulating GHGs under any provision of the CAA "could result in an unprecedented expansion of EPA authority that would have a profound effect on virtually every sector of the economy...."[54] Under the Obama Administration, EPA is moving toward the first option listed by the Supreme Court—an "endangerment finding"—by mid-April, 2009, followed by a 60-day comment period before the proposed finding is finalized.

As the ANPR asserts, the Court's ruling in *Massachusetts v. EPA* has many implications beyond its four corners.

On the substantive (non-standing) side, the Court's ruling upholding CAA coverage of GHG emissions from mobile sources improves the prospects of litigation seeking to have EPA restrict GHG emissions from stationary sources as well. The stationary-source provisions of the CAA use terms similar to that of section 202—in particular, "air pollutant," "in his judgment," and "endanger."[55] As the earlier subsection on suits seeking general CAA rulemaking indicated, such an effort to compel EPA regulation of stationary source GHGs is already underway as to NSPSs. Further, if EPA sets a national ambient air quality standard for CO2, GHGs would be covered under the CAA's new source review permitting program for major emitting facilities and modifications in Prevention of Significant Deterioration areas.[56] Presumably, best available control technology for CO2 emissions would then have to be installed on such facilities.[57]

On the mobile-source side, *Massachusetts v. EPA* is expressly relied upon in at least seven additional rulemaking petitions seeking EPA regulation of GHGs from mobile sources. As described in the ANPR, the petitions seek rulemaking under CAA sections 202(a)(3), 211, 213, and 231 to limit GHG emissions from (1) fuels and a wide array of mobile sources including oceangoing vessels, (2) all other types of nonroad engines and equipment, such as locomotives, construction equipment, farm tractors, forklifts, harbor crafts, and law and garden equipment, (3) aircraft, and (4) rebuilt heavy-duty highway engines.[58]

Beyond the federal clean air program, the Supreme Court's decision will likely be pivotal to the fortunes of plaintiffs in other climate change litigation owing to its discussion of standing. The question will be the extent to which the Court's finding of standing was contingent, as it obliquely suggested, on the existence of a state-sovereign plaintiff[59] and the presence in the CAA of an explicit provision allowing the filing of administrative petitions.

Ironically, the "environmental win" in *Massachusetts v. EPA* has thwarted the environmental side in a climate-change-related nuisance case. One court used the decision as peripheral support for dismissing a nuisance action on "political question" grounds, reasoning that the Supreme Court has now found authority over GHG emissions to reside in the Federal Government.[60] In the future, the decision may also undermine federal common law claims, on the argument that Congress intended to leave no room for courts to develop overlapping federal common law restricting GHG emissions.

II. WILDLIFE STATUTES

Marine Mammal Protection Act

The Marine Mammal Protection Act (MMPA)[61] bars the taking of marine mammals, with exceptions. One exception is for "incidental takings" by specified activities.[62] It provides that persons "engage[d] in a specified activity (other than commercial fishing) within a specified geographical region" may request the Secretary of the Interior or Commerce to authorize, for up to five years, the incidental, but not intentional, taking of small numbers of marine mammals. The Secretary must grant the authorization if he/she makes certain findings—including that the effect of the incidental take will be "negligible"—and promulgates rules setting out permissible methods of taking by the specified activity.

In *Center for Biological Diversity v. Kempthorne*, No. 3:07-CV-0141 (D. Alaska April 22, 2008), *transferred from* No. 07-CV-00894 (N.D. Cal. filed February 13, 2007), environmental groups challenge one such "incidental taking" rule -- authorizing the incidental take of polar bears and Pacific walrus for five years (2006-2011) resulting from oil and gas activities in the Beaufort Sea and adjacent coastal areas of the Alaska north slope.[63] Plaintiffs argue that the rule violates the MMPA by permitting more than a "negligible" impact on the species, based on the *combined* impact of oil-and-gas activities and the weakened condition of polar bears due to climate change.[64] The district court dismissed the suit, holding that the determination by the Fish and Wildlife Service (FWS) of negligible impact was reasonably based on the administrative record. An appeal has been filed. (This lawsuit also contains a National Environmental Policy Act claim, discussed in Section IV.)

Endangered Species Act

Under the Endangered Species Act (ESA),[65] animals (and plants) may be listed as endangered or threatened. Particularly relevant to climate change litigation are ESA sections 9 and 7.[66]

Section 9 makes it unlawful to "take" a member of a listed endangered species,[67] and has been extended by regulation to most threatened species.[68] Exceptions from the take prohibition are possible, chiefly through incidental take permits. The other provision, section 7, demands that each federal agency "insure that any action authorized, funded, or carried out by such agency ... is not likely to jeopardize the continued existence of any endangered species or threatened species or result in the destruction or adverse modification of [designated critical habitat] of such species...."[69] To achieve this goal, section 7 directs a federal agency to consult with the appropriate wildlife agency—the FWS or National Marine Fisheries Service (NMFS)—to determine the effect its action may have on listed species or their habitats. This is called "section 7 consultation." Then, the FWS or NMFS prepares a "biological opinion" concluding either that the proposed action would not violate the mandate of no jeopardy or adverse modification, or that it would violate the mandate, in which case FWS or NMFS must suggest "reasonable and prudent alternatives" that would not violate the mandate.

In *Natural Resources Defense Council v. Kempthorne*, 506 F. Supp. 2d 322 (E.D. Cal. 2007), environmental and sport fishing groups attacked the FWS biological opinion prepared for the 2004 Long-Term Central Valley Project and State Water Project Operations Criteria and Plan and certain related future actions. The biological opinion concluded that project operations would not jeopardize the continued existence of the Delta smelt, a threatened species, or adversely modify its designated critical habitat—that is, would not violate ESA section 7. The court, however, held that the biological opinion was arbitrary and capricious in ignoring data about climate change that may adversely affect the Delta smelt and its habitat. The court observed, for example, that the opinion was based on the assumption that the hydrology of the waters affected by the 2004 plan would follow historical patterns for the next 20 years, an assumption that studies on the potential effects of climate change on water supply reliability did not support.

A companion case pending before the same judge, *Pacific Coast Federation of Fishermen's Associations/Institute for Fisheries Resources v. Gutierrez*, No. 1:06-CV-00245, 2008 WL 2223070 (E.D.Cal. May 20, 2008), successfully challenged the NMFS biological opinion prepared in connection with the same project for various salmon and trout species—based on its "total failure to address, adequately explain, and analyze the effects of global climate change on the species." Id. at *60.

More ESA cases are likely on the way in connection with a campaign spearheaded by the Center for Biological Diversity (CBD). CBD has filed multiple petitions to have animals listed as endangered or threatened due in various degrees to climate change impacts on their habitat. Given that some of these petitions have been successful (and more may be in the future), the Center is likely to test in court whether substantial GHG sources run afoul of protections afforded those species by the ESA.

Three climate-change-related proposals to list a species have reached the actual listing stage thus far. The first, in which climate change is only a contributing factor, was NMFS' listing of the staghorn coral and elkhorn coral as threatened in 2006.[70] The second, garnering considerably more attention, was the May 15, 2008 listing of the polar bear as threatened,[71] under pressure of a court-imposed deadline requiring a decision for or against listing by that date.[72] The polar bear listing was based largely on the many studies as to the disproportionately large impact of climate change on the Arctic and the resulting loss of sea ice required by polar bears as habitat.[73] The third, again with climate change but a contributing factor, is NMFS' listing of the black abalone as endangered in 2009.[74] In addition to the coral, polar bear, and abalone, CBD has petitioned the FWS to list as either endangered or threatened Kittlitz's murrelet, a seabird (2001), 12 species of penguins (2006), the American pika, an alpine mammal (2007), the ashy storm-petrel, another seabird (2007), the ribbon seal (2007),[75] the Pacific walrus (2008), and the ringed, bearded, and spotted seals (2008). In each instance, the Center asserts global warming to be a cause, principal or otherwise, of the species' plight. (Not included in this report are the CBD suits challenging agency failures to make the statutorily mandated interim findings in the petition process for listing, known as 90-day or 12-month findings.)

With the listing of the corals and polar bear—particularly the latter where the climate change nexus is so clear—the question moves to the fore whether operating a fossil-fuel-fired power plant or other major GHG source violates section 9—causes a prohibited "take"—through the effects of its GHG emissions, via climate change, on polar bear habitat.[76] Notable here is that "take" is statutorily defined to include "harm" to a member of a listed species, and "harm," in turn, is defined by regulation to include certain "significant habitat modification[s] or degradation[s]."[77] The crux, presumably, is whether the causal link between the power plant's GHG emissions and the effect on the species habitat is sufficiently direct and substantial to constitute a "take," a question beyond the scope of this report. If a take is found, the power plant would require an incidental take permit to operate, such permit likely containing restrictions on the amount of GHGs that could be emitted. Likewise, the argument runs, a federal agency issuing a permit for power plant construction might have to initiate section 7 consultation.

In 2008, under the George W. Bush Administration, the FWS repeatedly asserted that its listing of the polar bear would not implicate the ESA—neither section 9 nor section 7 -- based on the GHG emissions from an activity. The FWS sought to ensure the irrelevance of GHG emissions to the ESA in several ways. One way was by issuing a "special rule" for the polar bear under ESA section 4(d) stating that section 9 "take" prohibitions do not apply to "any taking of polar bears that is incidental to, but not the purpose of, carrying out an otherwise lawful activity" occurring anywhere in the United States except Alaska.[78] A half-dozen or more lawsuits challenging the polar bear listing and the accompanying "special rule"—most including grounds related to climate change -- were consolidated on December 3, 2008 in the D.C. federal district court by the Judicial Panel on Multidistrict Litigation.[79]

Another way used by FWS (and NMFS) to keep the ESA and GHG emissions separate was by amending the section 7 consultation regulations to say that no consultation is required when a federal agency action is not anticipated to result in "take" and the action's effects are "manifested through global processes" and either (a) cannot be reliably predicted at the scale of the species' range, or (b) will have insignificant impact on the species or its habitat.[80] The amended section 7 regulations also lessen the chance that GHG emissions will trigger consultation by defining "indirect effects" of federal agency actions narrowly.[81] Owing in greater or lesser degree to the amended rule's impact on section 7 consideration of climate change, three lawsuits challenging the rule have been filed by environmental groups in the federal district court for the Northern District of California,[82] and one has been filed there by the State of California.[83] They will likely be consolidated.[84]

With the arrival of the Obama Administration, Congress in 2009 enacted a provision stating that the relevant Secretary may withdraw the polar bear special rule and the 2008 amendments to the consultation regulations "without regard to any provision of statute or regulation that establishes a requirement for such withdrawal." This streamlined withdrawal authority expires 60 days from March 11, 2009.[85]

III. ENERGY STATUTES

Energy Policy and Conservation Act

In *Center for Biological Diversity v. National Highway Traffic Safety Administration*, 538 F.3d 1172 (9th Cir. 2008), 11 states (CA, CT, ME, MA, NJ, NM, NY, OR, RI, VT, MN), environmental groups and others attacked a 2006 rule promulgated by the National Highway Traffic Safety Administration (NHTSA) under the Energy Policy and Conservation Act (EPCA). The rule established corporate average fuel economy (CAFE) standards for light-duty trucks—defined by NHTSA to include many SUVs, vans, and pickup trucks—in model years 2008 through 2011.

EPCA says that the light-truck CAFE standard shall be the "maximum feasible" standard that manufacturers can achieve in a given model year.[86] The court found that even assuming NHTSA may use a cost-benefit analysis to determine the "maximum feasible" standard, it was arbitrary and capricious not to include in the analysis the benefit of carbon emissions reduction—calling this "the most significant benefit of more stringent CAFE standards."[87] NHTSA had argued, for example, that the wide range of values put forward in studies as to how the benefits of reduced GHG emissions should be monetized justified placing no value on that benefit in its cost-benefit analysis. The court countered that while there is indeed a range of values in the studies, they are all greater than zero. Accordingly, the court remanded the CAFE standard to NHTSA for the agency to include a monetized value for carbon emission reduction in its analysis of the proper CAFE standard. (There was also a climate-change-related NEPA claim in this lawsuit, discussed in Section IV.)

Quite recently, CBD filed a petition for review of NHTSA's rule setting the standard for model year 2011 passenger cars and light trucks.[88] It is unclear from the tersely worded petition whether climate change concerns underlie this suit, though given the court decision immediately above, it seems likely.

Outer Continental Shelf Lands Act

In a petition for review, CBD challenges the June, 2007 approval by the Secretary of the Interior of the Outer Continental Shelf Oil and Gas Leasing Program 2007-2012. *Center for Biological Diversity v. U.S. Dep't of Interior* [sic], No. 07-1247 (D.C. Cir. filed July 2, 2007). CBD alleges that the Secretary violated the Outer Continental Shelf Lands Act[89] by failing to disclose or analyze the environmental and economic impacts from "the greenhouse gas emissions that would result from use of oil and gas produced as a result of the [Program]."[90] Note that it is not the GHG emissions from the oil and gas production itself that is at issue, but rather the GHG emissions resulting from the "use" of that oil and gas in cars, powerplants, or wherever. The defendant and intervenor-defendant briefs in this case focus heavily on standing, arguing among other things that only states have standing under *Massachusetts v. EPA*. (There was also a climate-changerelated NEPA claim in this lawsuit, discussed in Section IV.)

IV. INFORMATION STATUTES

Much of the climate change litigation is based on statutory requirements that the government generate, compile, or disclose information.

National Environmental Policy Act

To be sure, the National Environmental Policy Act (NEPA) is more than just an information statute, declaring as it does a sweeping policy that the federal government must consider the environmental impacts of its actions. However, NEPA ensures that such environmental consideration will occur chiefly through the production of information, in the form of environmental assessments and environmental impact statements, and does not require that an agency choose from among its options the one with the least environmental impact.

The NEPA cases involving climate change represent the oldest and most numerous category of climate change litigation. Again, not all cases are mentioned in this report.[91]

The dominant issue has been whether plaintiffs have standing to sue—as mentioned, an issue on which plaintiffs may be helped by the 2007 Supreme Court decision in *Massachusetts v. EPA*. Thus, all the standing issues discussed here should be viewed through the prism of that decision. The standing determination has been particularly difficult in the context of NEPA, which confers only a *procedural* right (having a federal agency prepare an adequate environmental impact statement (EIS)), not a *substantive* right (having the agency select a particular course of action after preparing the EIS). Where courts have found standing and reached the merits, they have usually accepted that climate change impacts in the proper circumstances are a required consideration in an EIS.[92]

District of Columbia Circuit

Standing barriers have proved particularly daunting in the D.C. Circuit, thus it is here that *Massachusetts v. EPA* may have its greatest effect. In the first significant climate change case, *City of Los Angeles v. National Highway Traffic Safety Admin.*, 912 F.3d 478 (D.C. Cir. 1990), the city attacked a NHTSA decision not to prepare an EIS when it set the corporate average fuel economy standard at 26.5 mpg for model year 1989 passenger cars—below the statutory default setting of 27.5 mpg.[93] A majority of the D.C. Circuit panel concluded that petitioners had standing based on their argument that had NHTSA done an EIS considering the climate change impacts of its one mpg rollback, the agency might have rejected it. This provided the requisite causal nexus, said the majority, between NHTSA's decision not to do an EIS and climate change. In dissent, however, one judge argued that Article III demanded a more precise causal showing, with clear proof of a nexus between the agency action and harm to the petitioners. On the merits, one judge in the majority concluded that NHTSA had "inadequately explained why the admitted increase in carbon dioxide is insignificant within the context of the environmental harm posed by global warming."[94] She would have remanded NHTSA's NEPA decision but left the rollback in place in the meantime. Because the other majority judge ruled for the agency, however, the petition was denied.

The plaintiff-friendly *City of Los Angeles* standard for finding global-warming-based standing was to prove short-lived. Six years later, a divided D.C. Circuit declared *en banc* that to obtain standing, a procedural-rights plaintiff must show not only that the government omitted a required procedure, but that it is substantially probable that the procedural omission *will cause a particularized injury to the plaintiff*[95]—adopting the dissenter's position in that case. To the extent *City of Los Angeles* dispensed with the second, causation-of-a-particularized-plaintiff-injury requirement, it was expressly overruled. Still later court decisions, however, have cast doubt on this strict standard.[96]

In *Foundation on Economic Trends v. Watkins*, 794 F. Supp. 395 (D.D.C. 1992), the standing bar was raised during, rather than after, the litigation. Plaintiffs claimed that NEPA required the Secretaries of Energy, Agriculture, and the Interior to evaluate the effect on climate change of 42 actions and programs under their authority. Plaintiffs' standing argument was based on "informational standing," under which failure to do an EIS discussing possible climate change impacts satisfies the injury requirement of standing merely by harming plaintiffs' programs for disseminating information about climate change to the public. In so arguing, plaintiffs relied on a line of D.C. Circuit decisions going back two decades. Unfortunately for them, however, informational standing was limited by the D.C. Circuit during the pendency of their suit. An amended complaint by the individual plaintiff, arguing that his expected use of his oceanfront cottage may be curtailed if oceans rise from climate change, was also rejected. Among other things, said the court, the plaintiff had not met the causation requirement of standing in that he had not related the environmental harm he predicted to any of the 42 challenged agency actions. "[T]here is no 'global warming' exception to the standing requirements of Article III or the [Administrative Procedure Act],"[97] it asserted.

In a suit described in Section III, *Center for Biological Diversity v. U.S. Dep't of Interior* [sic], No. 07-1247 (D.C. Cir. filed July 2, 2007), plaintiff charges that the Secretary of the Interior failed to analyze in the EIS for his five-year Outer Continental Shelf leasing program (1) the GHG emissions resulting from the use of the oil and gas produced under the program, and (2) the effects of global warming on the resources affected by the program "including, but not limited to, polar bears, walrus, and corals."

In *Montana Environmental Information Center v. Johanns*, No. 07-CV-1311 (D.D.C. filed July 23, 2007), *dismissed* March 20, 2008, challenge was made to the Department of Agriculture's Rural Utility Service's use of low-interest loans to help finance the construction of at least eight new coal-fired powerplants. The charge was that the EIS for one plant is deficient because it fails to consider the cumulative impacts of GHG emissions from the eight new plants.

Ninth Circuit

The standing barriers in the D.C. Circuit seem to be attenuated in the Ninth Circuit where, as far as research reveals, plaintiffs raising climate change claims in NEPA suits have yet to encounter standing problems.

In 2002, environmental groups sued the Overseas Private Investment Corp. (OPIC) and Export-Import Bank of the United States alleging continued failure to comply with NEPA. These federal agencies provide insurance, loans, and loan guarantees for overseas projects, or to U.S. companies that invest in overseas projects. Plaintiffs alleged that these overseas projects include oil and gas extraction and refining, and power plants, which together result in the annual emission of billions of tons of GHGs, causing climate change in the United States.

In 2005, the district court held that plaintiffs had standing, given what it saw to be the relaxed standards in the Ninth Circuit for showing standing in cases alleging procedural violations—here, failure to prepare an EIS.[98] *Friends of the Earth v. Mosbacher*, 2005 Westlaw 2035596 (N.D. Cal. 2005). It is "reasonably probable," said the court, that emissions from projects supported by the defendants will threaten plaintiffs' concrete interests. In 2007, the court reached the merits, holding on summary judgment motions that defendants need not prepare a programmatic EIS for the energy projects they finance, and that neither side had shown, as a matter of law, that energy projects specifically listed in the complaint are or are not "major Federal actions" requiring an EIS. 488 F. Supp. 2d 889 (N.D. Cal. 2007). The case was settled February 6, 2009, the Export-Import Bank and OPIC agreeing to implement various measures for considering the GHG emissions of supported projects.[99]

In *Border Power Plant Working Group v. Dep't of Energy*, 260 F. Supp. 2d 997 (S.D. Cal. 2003), plaintiff challenged the environmental assessment accompanying applications for permits and federal rights of way to build electricity transmission lines connecting new power plants in Mexico with the power grid in Southern California. In part because four of its members were seen to have procedural standing, the plaintiff organization was held to have organizational standing.[100] The court's standing discussion made no mention of climate change, however, perhaps because climate change was only a small part of plaintiff's case. On the merits, the court agreed with plaintiff that the environmental assessment was legally inadequate because, among other things, it failed to discuss CO_2 emissions from the powerplants and "[t]he record shows that carbon dioxide ... is a greenhouse gas."[101]

The decision in *Center for Biological Diversity v. NHTSA*, 538 F.3d 1172 (9th Cir. 2008), offers a deja vu to *City of Los Angeles*, discussed earlier in this section. Both cases involve a NHTSA rule setting a corporate average fuel economy (CAFE) standard—this time, for light-duty trucks (model years 2008-2011)[102]—and in both cases, the agency did no EIS. Petitioners include 11 states (CA, CT, ME, MA, NJ, NM, NY, OR, RI, VT, MN) and four environmental groups. In sharp contrast with earlier NEPA/climate-change decisions, the United States in this case did not contest standing and the court decision does not mention it.

On the merits, the court held that NHTSA's environmental assessment for its CAFE rule, finding no significant impact, was inadequate owing to, among other things, its analysis of the rule's cumulative impacts from GHG emissions. Said the court: "The impact of greenhouse gas emissions on climate change is precisely the kind of cumulative impacts analysis that NEPA requires agencies to conduct."[103] Nor did the Energy Policy and Conservation Act, the statute authorizing CAFE standards, limit NHTSA's duty to assess environmental impacts such as climate change. More specifically, while NHTSA's assessment indicated the expected amount of CO_2 emitted by light-duty trucks under the new CAFE standard, it failed to "evaluate the 'incremental impact' that these emissions will have on climate change ... in light of other past, present, and reasonably foreseeable actions such as other light truck and passenger automobile CAFE standards."[104] Finally, the court invoked the well-settled principle that an EIS must be prepared if substantial questions are raised as to whether a proposed project *may* have significant environmental impact, and held that petitioners' evidence raised the necessary level of doubt. Thus, the court ordered preparation of a full EIS. (There was also a climate change-related Energy Policy and Conservation Act claim, discussed in Section III.)

In *Center for Biological Diversity v. Kempthorne*, No. 3:07-CV-0141 (D. Alaska), *transferred from* No. 07-CV-00894 (N.D. Cal. filed February 13, 2007), environmental groups challenge a Fish and Wildlife Service "incidental taking" rule. As described in Section II, the rule authorizes the incidental take of polar bears and Pacific walrus by oil and gas activities in the Beaufort Sea and adjacent coastal areas of the Alaska north slope, from 2006 to 2011.[105] Plaintiffs challenge the environmental assessment and finding of no significant impact, charging that the Service put out the rule "without seriously analyzing the effects of climate change on them or their habitat." The accusation is not that the oil and gas activities themselves contribute to climate change, but that direct harms to polar bears and walruses from those activities will be exacerbated by climate change impacts on the Arctic that are already stressing those species. In April, 2008, the district court ruled that the FWS had been reasonable in finding that the impacts of oil and gas activities in and along the Beaufort Sea, over the next five years, will fall short of NEPA's "significant" threshold for requiring environmental assessments. An appeal has been filed.

Eighth Cir cuit

In *Mid States Coalition for Progress v. Surface Transportation Bd.*, 345 F.3d 520 (8th Cir. 2003), petitioners disputed the adequacy of an EIS prepared by the Surface Transportation Board to accompany its approval of a railroad's proposal to construct new rail and upgrade existing rail. The proposed rail line was to provide a less expensive route by which low-sulfur coal in Wyoming's Powder River Basin could reach powerplants, and thus might be expected to increase coal consumption and its attendant effects. In this regard, the court noted that CEQ's NEPA regulations require that EISs cover both direct and indirect effects of proposed actions.[106] It concluded by finding it "irresponsible" for the Board to approve such a large project without first examining the possible effects of an increase in coal consumption— apparently, the opinion suggests (but does not explicitly say), including climate change.[107]

In *Ranchers Cattlemen Action Legal Fund v. Conner*, No. 07-CV-01023 (D.S.D. filed October 24, 2007), plaintiffs challenge Department of Agriculture regulations easing restrictions on the import of live cattle and edible bovine products from "minimal risk" Mad Cow Disease regions (Canada). Plaintiffs assert that the environmental assessment was inadequate because it did not analyze the increased GHG emissions from the transportation of the cattle into the United States.

State NEPAs

A few GHG-related suits also have been filed under state "little NEPAs"—state laws requiring state (and sometimes local) agencies to consider the environmental impacts of their proposed actions, just as the federal NEPA does for federal agencies.[108] For example, in *General Motors Corp. v. California Air Resources Bd.*, No. 05-02787 (Cal. Sup. Ct. filed September 2, 2005), two car manufacturers claimed that the Board's adoption of California's GHG emission standards involved delayed and inadequate compliance with the state's NEPA-type law. This suit offers as a prime reason for environmental analysis the argument that GHG emissions regulation has, in addition to a possible benefit, some environmental downsides. In particular, it contends that restriction of GHG emissions may cause an increase in new-vehicle sticker prices and a consequent decrease in the rate at which old, higher-emissions vehicles are retired from use.

Invoking California's NEPA-like statute (the California Environmental Quality Act), conservation groups and California attorney general Jerry Brown sued in 2007 to require San Bernardino County, the largest county in the US by area, to address climate change in its General Plan update.[109] Later that year, California settled its lawsuit, the county agreeing to prepare a Greenhouse Gas Emissions Reduction Plan and adopt other measures.[110] Later, the conservation groups took a voluntary dismissal of their suit after extracting promises from the county to do a mapping of wildlife habitat and research on wildfire dangers. In broaching the vast realm of local land use plans, these cases portend a major new front in climate change litigation, particularly in states that require environmental impact analysis.

Global Change Research Act

The Global Change Research Act of 1990 (GCRA)[111] commands the President to create an interagency United States Global Change Research Program to improve understanding of "global change." Global change is defined broadly by the GCRA to include all changes in the global environment "that may alter the capacity of the Earth to sustain life." Thus, the statute includes, but goes beyond, climate change.[112] The Program is to be implemented by a National Global Change Research Plan, with regular scientific assessments that evaluate the findings of the Program. The GCRA demands that revised Research Plans be submitted to the Congress at least every three years,[113] with the last one having been submitted July, 2003. The statute further demands that scientific assessments be submitted to the President and Congress not less often than every four years,[114] with the only assessment to date submitted October, 2000.

On these undisputed facts, the district court in *Center for Biological Diversity v. Brennan*, 571 F. Supp. 2d 1105 (N.D. Cal. 2007), had little difficulty finding that the Bush Administration had unlawfully withheld action it was required to take. It ordered defendants to publish a summary of the revised proposed Research Plan no later than March, 2008, with submission to Congress no later than 90 days thereafter.[115] The court further ordered the scientific assessment to be produced by May, 2008. It should be noted that the great bulk of this opinion is devoted not to the foregoing violation and remedy, but to threshold matters: standing (finding procedural rights injury and informational injury, associational standing, and Administrative Procedure Act standing) and a motion to intervene by two Members of Congress (denied).

Freedom of Information Act

The Freedom of Information Act (FOIA)[116] mandates that documents in the possession of federal executive branch agencies are to be disclosed to the public upon request, unless covered by a FOIA exemption.

In May, 2006, Citizens for Responsibility and Ethics in Washington (CREW) invoked FOIA to request from the Council on Environmental Quality (CEQ) all of its records relating to the causes of climate change, from January 20, 2001, to October 26, 2006. Though CEQ produced many documents, CREW sued under FOIA seeking a court order that CEQ release all records responsive to its request. *Citizens for Responsibility and Ethics in Washington v. Council on Environmental Quality*, No. 1:07CV00365 (D.D.C. filed February 20, 2007). The case has been stayed while CEQ efforts to comply continue.

This lawsuit parallels allegations at the time that political appointees in the Bush Administration CEQ edited many of the agency's reports to minimize the danger and human causes of climate change. In July, 2006, the House Committee on Government Reform[117] requested that CEQ provide documents and communications relating to the agency's edits of climate change materials, its efforts to influence the statements of government scientists, its communications with federal agencies and nongovernmental parties regarding climate change, and so on. A report making findings was issued in December, 2007,[118] with minority views.[119]

V. COMMON LAW TORT

The widely diverse injuries predicted from climate change mean that a comparably diverse spectrum of plaintiffs and defendants could become involved in common law tort litigation based on such injuries. Possible plaintiffs include property owners (farmers dealing with reduced rainfall, owners of oceanfront homes dealing with rising sea level or increased storm activity), nonowner users of natural resources (ski resort operators, commercial fishermen), and state attorneys general bringing private or public nuisance claims (the former for injury to state-owned land, the latter on behalf of the state's citizenry to protect public health and well-being). Possible defendants include the companies that produce the fossil fuels whose combustion produces GHG emissions, entities that emit GHGs (chiefly fossil-fuel-fired powerplants, but many other sources also), and companies that manufacture or market products whose use creates GHG emissions (chiefly the automakers).[120]

Several of these potential plaintiff and defendant categories are represented in the five climatechange-related tort cases known to be filed thus far (four discussed in the following text, one in footnote 130). Thus far, all of those tort actions that have produced court decisions have failed, either due to lack of standing or the political question doctrine, or both. Three are on appeal, however.

Nuisance

Nuisance has been the principal tort theory used in cases seeking relief (injunctive or monetary) from harms caused by climate change.[121]

In 2004, eight states (CA, CT, IA, NJ, NY, RI, VT, WI) and New York City sued five electric utility companies.[122] *Connecticut v. American Electric Power Co.*, Civ. No. 04 CV 05669 (S.D.N.Y. filed July 21, 2004). These defendants were chosen as allegedly the five largest CO2 emitters in the United States, through their fossil-fuel-fired electric powerplants. Invoking the federal and state common law of public nuisance,[123] plaintiffs seek an injunction requiring defendants to abate their contribution to the nuisance of climate change by capping CO2 emissions and then reducing them. Plaintiffs sue both on their own behalf to protect state-owned property (e.g., the hardwood forests of the Adirondack Park in New York), and as *parens patriae* on behalf of their citizens and residents to protect public health and well-being.

On the same day, three land trusts filed a similar suit against the same defendants, in the same court, adding a private nuisance claim.[124] *Open Space Institute v. American Electric Power Co.*, No. 04 CV 05670 (S.D.N.Y. filed July 21, 2004). They seek to protect land owned and preserved by them in the state of New York, which they claim to be threatened by climate change.[125] This suit was consolidated with the state suit.

In a series of motions, defendants sought to have these actions dismissed on a wide spectrum of threshold grounds. Though the case has now been decided by the trial court on a single threshold issue, it is worth reviewing some of the grounds advanced in these motions because they may reappear later, in this or other private GHG litigation. To reiterate, many of these grounds typify the difficulties encountered when one seeks to address through private litigation a ubiquitous, long-term environmental problem to which countless parties contribute.

In a dismissal motion, some defendants argued there is no federal common law cause of action for climate change. Creating such federal common law, they argued, runs afoul of Supreme Court directives that federal courts do so only in limited areas—especially where, as with climate change, the problem at issue has sweeping implications. Even assuming a viable federal common-law nuisance theory, they continued, Congress's enactment of a comprehensive scheme of air pollution regulation in the CAA displaces federal court authority in this area. Defendants also challenge plaintiffs' standing to sue. Plaintiffs, they argued, have not demonstrated the "injury in fact" requisite of standing because they allege only injuries from climate change in the indefinite future. Nor, said these defendants, have plaintiffs shown "causation" because they do not allege that defendants' conduct will directly cause the consequences of climate change—especially since defendants' collective emissions are admitted to be less than 2-1/2% of the global total from human activities.[126] As mentioned, the viability of these federal common law of nuisance and no-standing arguments by defendants may be significantly affected—the displacement argument helped, the others hurt—by *Massachusetts v. EPA*.

Another motion to dismiss asserted that to the limited extent a federal common law claim to abate an interstate nuisance has been recognized, it has been limited to actions brought by state entities. Nor, said defendants, can plaintiffs assert public nuisance, because they have not alleged special injury to their properties, or private nuisance, because they have not alleged substantial harm.

As indicated, the dismissal motions in *Connecticut* and *Open Space Institute* have now been ruled on by the district court,[127] which dismissed the cases on political question grounds. This judicial doctrine requires a court to look into "the appropriateness under our system of government of attributing finality to the action of the political departments [i.e., the legislative and executive branches] and also the lack of satisfactory criteria for a judicial determination...."[128] One situation judicially recognized as pointing to a political question, hence dismissal of the action, is "the impossibility of deciding [the case] without an initial policy determination of a kind clearly for nonjudicial discretion."[129] This situation, said the court, perfectly fit the GHG cases, touching as they do on so many areas of national and international policy. As a political question, the court believed the climate change issue in these suits to be for the legislature, not the courts, to resolve. Very possibly, the amorphousness of nuisance law, giving the court little guidance in resolving these cases, may have hurt the plaintiffs' cause. These cases are now on appeal to the Second Circuit.[130]

A second nuisance action was filed in 2006 by California against several automobile manufacturers based on the alleged contributions of their vehicles to climate change impacts in the state. The suit asserts that these impacts constitute a public nuisance under federal or state common law, and seeks monetary damages (plaintiffs in *Connecticut* seek injunctive relief). The district court dismissed the suit on the same political-question rationale as in *Connecticut*— namely, "the impossibility of deciding without an initial policy determination of a kind clearly for nonjudicial discretion." *California v. General Motors Corp.*, 2007 Westlaw 2726871 (N.D. Cal.

September 17, 2007). The need for an "initial policy determination" by the political branches was supported, in the court's view, by the complexity of the climate change issue, the need for political guidance in divining what is an "unreasonable" interference with the public's rights (the definition of a public nuisance), and the global warming debate in the political branches. Ironically, the environmental "win" in *Massachusetts v. EPA* was cited by the court against the state, both because that decision found authority over GHG emissions to lie with the federal government and because it recognized a state's standing to press its grievances at the federal level. An appeal to the Ninth Circuit is pending.

Most recently, a native village on the northwest Alaska coast sued certain oil and energy companies, claiming that the large quantities of GHGs they emit collectively contribute to climate change. Climate change, the village contends, is destroying the village by melting Arctic sea ice that formerly protected it from winter storms, leading to massive coastal erosion. *Native Village of Kivalina v. Exxonmobil Corp.*, No. 08-cv-01138 (N.D. Cal. filed February 26, 2008). Indeed, the complaint asserts, "[t]he U.S. Army Corps of Engineers and U.S. Government Accountability Office have both concluded that the village must be relocated due to global warming...." The village invokes the federal common law of public nuisance, and state statutory or common law of private and public nuisance, and makes a civil conspiracy claim. The conspiracy claim asserts that some of the defendants have engaged in agreements to participate in the intentional creation or maintenance of a public nuisance—that is, global warming—by misleading the public as to the science of global warming. The suit seeks monetary damages.

Negligence, etc.

Owners of Mississippi property damaged by Hurricane Katrina sued certain oil, coal, and chemical companies, alleging a multistep chain of causation: the companies emitted GHGs, which contributed to global warming, which made the waters of the Gulf of Mexico warmer, which caused Hurricane Katrina to become more intense as it passed over the Gulf than it would otherwise have been, which increased the harm to plaintiffs' property caused by the hurricane. Plaintiffs asserted various state-law tort claims, including negligence, nuisance (public and private), and trespass, and seek compensatory damages; they request punitive damages for gross negligence. Further, they claimed fraud and conspiracy to commit fraud, alleging that the oil and coal companies disseminated misinformation about global warming. Finally, plaintiffs made claims against their home insurance companies (e.g., breach of fiduciary duty claim for misrepresenting policy coverage, and violation of a state consumer-protection act) and their mortgage companies (arguing that they may not claim sums owed by plaintiffs for the value of the mortgaged property that was uninsured).

The district court, sitting in diversity, dismissed the action for lack of plaintiff standing. *Comer v. Murphy Oil USA, Inc.*, Civ. Action No. 1:05-CV-436-LG-RHW (S.D. Miss. August 30, 2007). With regard to certain defendants, the court also found plaintiffs' claims nonjusticiable under the political question doctrine—as in the decisions above where nuisance was the sole legal theory advanced. An appeal to the Fifth Circuit is pending.

VI. FEDERAL PREEMPTION

Stationary Sources of GHG Emissions

The question of whether federal law preempts state regulation of GHG emissions arises chiefly in connection with *mobile* sources. With limited exceptions, the CAA disclaims any intention to preempt state air pollution controls on *stationary* sources.[131] And the Energy Policy and Conservation Act does not set fuel economy standards for other than mobile sources, so it too would be unlikely to preempt state regulation of stationary sources. However, some have asserted that state regulation of stationary-source GHGs is preempted as contrary to the federal government's authority over foreign policy—an argument being pressed, so far unsuccessfully, in litigation attacking state regulation of *mobile-source* GHG emissions (see below). The most prominent state enactment limiting GHG emissions from stationary sources is that of California, which as yet has not been challenged.[132]

Mobile Sources of GHG Emissions: CAA Preemption

The picture is quite different for mobile sources, where preemption is the general rule. The CAA preempts states from adopting any "standard relating to the control of emissions from new motor vehicles ...,"[133] and the act defines "emission standard" as certain limits on "emissions of *air pollutants*."[134] The Supreme Court has now held that at least for purposes of mobile sources, "air pollutants" includes GHGs. Thus, CAA preemption of state regulation of car and truck GHG emissions is clear, whether or not EPA proceeds to regulate a particular mobile-source GHG. It would seem, then, that states are preempted from setting emission standards for CO2, methane, and hydrofluorocarbons— three substances said to enhance climate change—even though EPA has not set mobile-source emission standards for them.

An exception to the general CAA rule preempting state mobile-source emission regulation is that EPA may waive CAA preemption for one particular state, California, if that state requests a waiver.[135] Further, when EPA does grant California a waiver, the act automatically extends it to almost all states with mobile-source emission limits identical to California's.[136]

Under this "California waiver" authority, California requested a preemption waiver for its GHG emissions regulations on December 21, 2005. These regulations had been promulgated under a 2002 California enactment that was the first in the nation to call for limits on GHG emissions from mobile sources. Assembly Bill 1493[137] instructs the California Air Resources Board (CARB) to adopt regulations that achieve the maximum feasible reduction of GHGs emitted by passenger vehicles and light-duty trucks. The CARB adopted the required regulations in 2004. The regulations target CO2, methane, nitrous oxide, and hydrofluorocarbon emissions, setting "fleet average greenhouse gas exhaust mass emission requirements for passenger car, light-duty truck, and medium-duty passenger vehicle weight classes." The first model year to which the fleet averages apply is 2009. The averages are reduced for each subsequent model year through 2016.

On December 19, 2007, almost two years after California requested the waiver, the EPA Administrator wrote the California governor that he intended to deny the state's request. On January 3, 2008, two petitions for review of this letter, arguing that it constituted final agency action on the waiver request, were filed in the Ninth Circuit. However, with the issuance of EPA's March 6 decision document, [138] these suits based on the EPA letter were dismissed and replaced by a suit in the D.C. Circuit challenging that document. Petitioners in *State of California v. U.S. EPA*, No. 08-1178 (D.C. Cir. filed May 5, 2008) are California, 18 other states, and numerous environmental groups. Most of the California congressional delegation, including Speaker of the House Nancy Pelosi and Senators Boxer and Feinstein, are participating as amici in support of the petitioners.[139] With the arrival of President Obama, the California Air Resources Board and President Obama (by executive order) requested EPA to reopen the waiver-denial matter—which EPA did on February 12, 2009.[140] On February 25, 2009, motion was granted to hold *State of California* in abeyance pending the Obama Administration EPA's reconsideration of California's petition.[141]

Mobile Sources of GHG Emissions: Non-CAA Preemption

That the CAA preempts state GHG regulation of mobile sources cannot be seriously questioned, absent a California waiver. The following preemption litigation is significant for the *non-CAA* preemption claims being pressed. If successful, these claims would prevent California and other states from implementing the California mobile-source standards *even if EPA's denial of the waiver is administratively or judicially reversed.*

The chief non-CAA preemption theory in this litigation is based on the Energy Policy and Conservation Act (EPCA, also noted in Section III). EPCA is the authority under which the National Highway Traffic Safety Administration (NHTSA) establishes corporate average fuel economy standards ("CAFE standards").[142] As recently amended, EPCA requires NHTSA to prescribe separate fuel economy standards for passenger and non-passenger automobiles beginning with model year 2011, to achieve a combined fuel economy average for model year 2020 of at least 35 miles per gallon.[143] More pertinent here, EPCA preempts states from adopting laws "related to" the federal fuel economy standards.[144] The auto industry argues that the only known way to reduce GHG emissions is to improve gas mileage, so that a state regulation of auto GHG emissions is a law "related to" the federal emission standard, hence invalid.

Non-CAA preemption suits, brought by auto interests, are pending in four of the federal judicial circuits containing a state that has adopted GHG controls on vehicles. Two decisions on the merits have been handed down, from Vermont (First Circuit) and California (Ninth Circuit). Both reject the preemption theories presented.

In the first to be decided, *Green Mountain Chrysler Plymouth Dodge v. Crombie*, 508 F. Supp. 2d 295 (D. Vt. 2007), the court ruled that the relationship between Vermont's California-identical GHG standards and EPCA was better analyzed as an interplay between two federal statutes (EPCA and the CAA) rather than as a federal preemption question. So viewing the matter, the court pointed out that NHTSA has consistently treated EPA-approved California mobile source emissions standards as constituting "other motor vehicle standards of the Government," which EPCA says NHTSA must consider when setting CAFE standards.[145] This suggests that EPCA was meant to coexist with the CAA, rather than supersede it. Moreover, noted the court, in a related context the Supreme Court's *Massachusetts v. EPA* decision saw the EPCA CAFE provisions as harmonious with the CAA.[146] Thus, the court found the relationship between the CAA waiver authority and the EPCA CAFE provisions to be one of overlap, but not conflict. Despite its conclusion that preemption doctrine did not apply, the court did a preemption analysis anyway, finding that Vermont's GHG standards were preempted neither by EPCA nor as an intrusion upon the foreign policy authority of the United States. An appeal is pending.

In the second decision, *Central Valley Chrysler-Jeep, Inc. v. Goldstene*, 529 F. Supp. 2d 1151 (E.D. Cal. 2007), a district court similarly rejected claims that California's regulation of GHG emissions from cars and trucks was precluded by EPCA, preempted by EPCA, or preempted as an intrusion of state law on federal authority to conduct foreign affairs. An appeal in this case is pending as well.

The legal theories pressed in the *Crombie* and *Goldstene* litigation are similar to those in two Rhode Island suits, consolidated as *Lincoln Dodge, Inc. v. Sullivan*, No. 1:06-CV-00070 (D.R.I. filed February 13, 2006), challenging that state's adoption of the California standards. Recently, the district court held that the claims of the auto manufacturers and trade associations in this case were barred by collateral estoppel, a legal doctrine that prohibits parties from relitigating issues they have already adjudicated, as these plaintiffs had done in *Crombie* and *Goldstene*. The Rhode Island auto dealers, by contrast, had themselves never raised the issues in the case and thus were held to be viable plaintiffs, allowing the case to proceed. In yet another preemption case, New Mexico's adoption of the California GHG standards has been challenged as preempted under EPCA in *Zangara Dodge, Inc. v. Curry*, No. 1:07-CV-01305 (D.N.M. filed December 27, 2007).

VII. STATE STATUTES

The first climate-change decision involving state statutes (other than nuisance statutes) appears to be *Matter of Quantification of Environmental Costs*, 578 N.W.2d 794 (Minn. App. 1998). This case involved a state law requiring the state's public utilities commission to determine environmental cost values for each method of energy generation, and to use those values in proceedings before the commission. The commission set values for six pollutants, including CO_2. Petitioners' challenge to the CO_2 value was rejected on the grounds that (a) notwithstanding the speculative nature of some of the data, the AUJ conducted a careful review based on sufficient evidence in the record, (b) the determination that CO_2 negatively affects the environment was proper,[147] and (c) the determination as to CO_2 value otherwise comported with the governing statute.

In 2000, the City of Seattle adopted a goal of meeting its electricity needs with "no net greenhouse gas emissions." To achieve this goal, the city ordered the city-owned electric utility to offset its GHG emissions by paying others to reduce their GHG emissions. The utility did so, largely through agreements paying other entities to use cleaner fuels. This made the utility (according to its press release) "the first large electric utility in the country to effectively eliminate its contribution of harmful greenhouse gas emissions." In *Okeson v. City of Seattle*, 150 P.3d 556 (Wash. 2007) (en banc), however, the utility's ratepayers argued that this offset arrangement was not authorized by the state's utility enabling act. The Washington Supreme Court agreed, explaining that the purchase of GHG offsets was not impliedly authorized by the enabling act in that the offset contracts were not proprietary because they were not part of the services for which ratepayers are billed, nor were they within the enabling act's purposes.

A pair of cases deals with permit applications by electric utilities seeking to build new facilities. In *In re Otter Tail Power Co.*, 744 N.W.2d 594 (S.D. 2008), environmental intervenors urged the South Dakota public utilities commission to deny a permit to build a coal-fired energy conversion facility, in light of the substantial CO_2 it would emit. Notwithstanding, the commission granted the permit, and the state supreme court sustained. The commission, it held, was not clearly erroneous in finding that the added CO_2 threatened no "serious" injury to the environment, the state's statutory standard. Deference to the commission was particularly appropriate, it said, because the CO_2 from the facility would increase national CO_2 emissions by only .07%, and neither Congress nor the state had chosen to regulate CO_2 emissions.

By contrast, the permit was ultimately denied in Kansas. After applying for a PSD construction permit[148] for two 700-megawatt coal-fired power plants, the Sunflower Electric Power Corp. initially received a favorable response from the state agency, which asserted it would not consider CO_2 in connection with the application owing to the national and international character of climate change. Later, however, the agency invoked a state law providing it with emergency powers when emissions present a substantial endangerment to the health of persons or the environment.[149] Using this authority, and specifically citing the large volume of CO_2 from the proposed plants, the agency denied the permit in 2007. Three times in 2008 and once in 2009 the Kansas legislature passed laws that would have required Sunflower's application to be evaluated without taking CO_2 emissions into account, but each was vetoed by Governor Sebelius. In response to the 2008 vetoes, Sunflower filed several suits now pending in state and federal court. In federal court, in *Sunflower Electric Power Corp. v. Sebelius*, No. 08-2575 (D. Kan. filed November 17, 2008), Sunflower alleges first that the permit denial violates equal protection because it prohibits CO_2 emissions from the proposed plants when the state has authorized, and continues to authorize, other CO_2 sources in Kansas. Second, Sunflower claims a violation of the Dormant Commerce Clause[150] in that the permit denial was allegedly motivated by the fact that much of the electricity to be generated by the proposed plants would be sold out of state.

VIII. INSURANCE POLICY LITIGATION

Research reveals only one lawsuit contesting insurance policy coverage of injuries or liability arising from climate change, though the future is likely to see more. One of the energy companies sued in tort by the Village of Kivalina (see Section V) is now being sued by the insurance company holding its commercial general liability policy. *Steadfast Insurance Co. v. The AES*

Corporation, No. 2008-858 (Va. County Ct. filed July 9, 2008). The insurance company seeks a declaratory judgment that, it hopes, will decree it is not obligated under the policy to provide either defense or indemnity coverage to the energy company in the litigation brought by the Village of Kivalina. The insurer's arguments are three: (1) the policy applies only to an "accident," which is not the basis of the suit against the energy company by the *Kivalina* plaintiffs; (2) the policies do not apply to injury that began before the earliest of the insurance policies (September 2003), which the injuries here did; and (3) all of the conditions for avoiding the policy's pollution exclusion have not been met (e.g., the pollution alleged by the *Kivalina* complaint was not unexpected).

More significant than the coverage of current liability and casualty policies is the long-term challenge posed by climate change to the insurance industry.[151]

IX. INTERNATIONAL LAW

Reports suggest that the successor to the Kyoto Protocol may contain provisions by which wealthy industrialized nations contribute to the adaptation costs of developing countries affected by climate change. Lurking in the background, however, is the question whether the major GHG emitting nations can be sued in international fora for the adverse effects of climate change.

Gauging the viability of such claims involves a good deal of guesswork, as they lie on the frontiers of international law. This report, concerned primarily with actually filed claims, notes only a few highlights, taken mostly from what appears to be the most pertinent article in the area.[152] The article suggests that the International Court of Justice (ICJ) might be one forum for resolution of climate change claims, with jurisdiction established through treaties that specifically provide for dispute resolution before the court. A problem with the ICJ approach is that the treaties most likely to be invoked are Friendship, Commerce, and Navigation or similar treaties, which focus on how each party *within its own country* treats the other country's nationals and property. A climate change suit, by contrast, likely would have an extraterritorial focus. Another ICJ possibility would be for the court to render an advisory opinion, at the request of a body authorized under the U.N. Charter to request one.

Other possibilities include voluntary submission of a climate change dispute to any of several international arbitral forums or resort to the specialized dispute resolution systems created under various treaties. An example of the latter, reportedly being actively considered, is a fisheries conservation agreement under the UN Law of the Sea Convention, presumably on the argument that increased ocean temperatures from climate change imperil certain fish stocks.[153]

Some principles that might be applied to a claim alleging GHG-caused injury might be taken from the international law of transboundary pollution. For example, the Restatement (Third) of Foreign Relations Law describes an international law principle under which a nation must "take such measures as may be necessary, to the extent practicable under the circumstances, to ensure that activities within its jurisdiction or control ... are conducted so as not to cause significant injury to the environment of another state...."[154] Similarly, the *Trail Smelter* arbitration decision, probably the seminal ruling on state liability for transboundary pollution, declared that "[a] State owes at all times a duty to protect other States against injurious acts by individuals from within its jurisdiction."[155] Of course, as with the domestic common law litigation described in Section V, daunting hurdles confront the claimant in making the link between climate change in general and specific environmental harms, and in apportioning how much of such harms are attributable to the charged party or parties, in this instance the United States.

Research reveals only one climate-change-related international law action filed against the United States. Not surprisingly, it was filed by a group based in the Arctic, where the temperature rise from climate change has been among the fastest. In 2005, the Chair of the Inuit Circumpolar Conference, on behalf of herself and all affected Inuit of the arctic regions of the United States and Canada, filed a petition against the United States with the Inter-American Commission on Human Rights, the investigative arm of the Organization of American States (OAS).[156] The petition alleged that the United States, through its failure to restrict its GHG emissions and the resultant climate change, has violated the Inuit's human rights—including their rights to their culture, to property, to the preservation of health, life, and physical integrity, and so on.[157] Inuit culture is described in the petition as "inseparable from the condition of [its] physical surroundings." Generally, the Inter-American Commission on Human Rights is empowered to recommend measures that contribute to human rights protection, request states in urgent cases to adopt specific precautionary measures to avoid serious harm to human rights, or submit cases to the Inter-American Court of Human Rights. The United States, however, has not accepted the jurisdiction of this court, so the Inuit petition sought only to have the Commission prepare a report declaring the responsibilities of the United States and recommending corrective measures.

In 2006, the Inuit petition was rejected, with no reasons given (as is customary for the Commission). However, at the request of petitioners the Commission held a hearing on March 1, 2007 on the generic issue of climate change and human rights. One may speculate that the Commission felt more comfortable with the hearing format than the petition because the former did not single out the United States. Or that the Commission was concerned the petition took it into a realm of global scale, orders of magnitude vaster than the typical human rights petition it receives.

In 2005-2006, five petitions were submitted to the Intergovernmental Committee for the Protection of the Cultural and Natural Heritage of Outstanding Universal Value (World Heritage Committee), part of UNESCO.[158] The petitions request that various designated World Heritage Sites be placed on the List of World Heritage in Danger[159] owing to alleged impacts of climate change. The sites covered by the petitions are Waterton-Glacier International Peace Park (U.S./Canada), Sagarmatha National Park (Nepal), Belize Barrier Reef Reserve System (Belize), Huascaran National Park (Peru), and the Great Barrier Reef (Australia). Only the Waterton-Glacier petition was filed by entities within the United States (12 environmental groups) and involves a natural resource within the United States. As a party to the World Heritage Convention, the United States is obligated to "do all it can ... to the utmost of its own resources and, where appropriate, with any international assistance and cooperation" to protect its cultural and natural heritage.[160]

In 2006, the World Heritage Committee acknowledged the five petitions but appeared desirous of shifting the debate toward the use of existing committee mechanisms at individual sites to adapt to the threat of climate change.[161] Since then, a few additional petitions to place sites on the danger list have been filed, most interestingly a petition titled "The Role of Black Carbon in Endangering Sites Threatened by Glacial Melt and Sea Level Rise."[162] This petition notes that "[r]ecent scientific studies identify black carbon, a component of fine particulate matter, as a key climate forcing agent...."[163] It then asserts that high latitude and high altitude glaciers and low-elevation sites are the World Heritage Sites most vulnerable to climate change, and lists 15 sites (including those in the preceding paragraph) that should be placed on the danger list.[164] Waterton-Glacier remains the only site mentioned in a petition for placement on the List of World Heritage in Danger that is in the United States.

Thus far, no international law claims have been brought by low-lying nations likely to be inundated by the sea level rise predicted to accompany climate change. A recent scientific report asserts that sea level rise is likely to be larger than previously predicted, affecting as many as 600 million people on low-lying Pacific islands and southeast Asia delta areas.[165]

X. COMMENTS

Gauging the prospects of the pending climate change lawsuits is a precarious venture; for many of the suits, there is little precedent. It is clear, however, that success in the conventional sense— obtaining a judgment for the environmental plaintiff—is not the only objective of many of these suits. Some of the climate change litigation almost certainly has a long-range strategic purpose— to keep climate change on the political front burner and make it difficult for government and GHG emitters to ignore the problem.

In the conventional sense of the term, plaintiffs' successes have been rare in cases seeking relief *directly from GHG emitters*. A court may be reluctant to impose expensive measures to address a global problem on a defendant that is a proportionately minor contributor (which almost all defendants are, given the vast number of GHG emitters), using statutory provisions or common law principles that were not formulated with global problems in mind, against a backdrop of scientific uncertainty as to the precise consequences (if not the general cause) of climate change. By contrast, the environmental side recently has scored major wins where *governmental* remedies were sought. In a string of 2007 decisions under the Clean Air Act,[166] Energy Policy and Conservation Act of 1975,[167] foreign policy authority of the United States,[168] and NEPA,[169] courts have shown increased willingness to authorize or require government consideration of climate change.

As this report shows, standing has been a persistent issue for environmental plaintiffs, though of late the tide appears to be shifting in their favor. And at least for states, the Supreme Court decision in *Massachusetts v. EPA* is likely to work a sea change in improving plaintiffs' prospects. As noted earlier, the big question is the extent to which the Supreme Court decision finding standing will be seen by the lower courts to generalize to non-state plaintiffs, other statutory and common law contexts, and injuries (as from weather events) not as clearly attributable to climate change as Massachusetts's loss of shore land.

Causation is not only a component of the threshold standing test but a component of the plaintiff's case on the merits. Several writers have identified proof of causation as a key obstacle to a tort action seeking relief from climate change injury.[170] And at the remedy stage, allocation of damages among specific defendants will likely present both factual and legal challenges.

In either the standing or case-in-chief contexts, the climate change issues in private-remedy actions reprise an intractable problem in environmental law: imposing liability for harms that are remote in time and place from the pollution sought to be abated, particularly where the pollution comes from multiple sources.[171] Lawmakers of yesteryear encountered this same redistributive conundrum in tackling the problem of acid rain, where pollution cause and effect are separated by hundreds of miles and weeks or months. Imposing liability for harm from exposure to toxic chemicals is of the same ilk: exposure to contamination from multiple sources may result in ill effects manifested only a decade or two later.

Perhaps because of these hurdles under existing law, and the resistance of the George W. Bush Administration to regulatory approaches to climate change, new directions have been explored.[172] Within the United States, several states have adopted their own GHG emission controls, citing, among other things, inaction at the federal level.[173] Twenty-three states have joined one of the three regional GHG reduction initiatives (Western Climate Initiative, Midwestern Regional Greenhouse Gas Reduction Accord, and in the northeastern states, Regional Greenhouse Gas Initiative).[174] Some states have explored the idea of emissions trading with Europe.[175] At least 40 states and multiple Canadian provinces have partnered to form a Climate Registry to support voluntary and mandatory schemes for reporting GHG emissions in those states and provinces. California and the United Kingdom signed an agreement on July 31, 2006, committing both parties to implement market-based mechanisms, to share results from studies to quantify the economic impacts of climate change, collaborate on research, etc.[176] Also internationally, this report noted the unsuccessful Inuit petition filed with the Inter-American Commission on Human Rights and the pending petitions before the World Heritage Committee. Reportedly, the low-lying Pacific nation of Tuvalu threatened to sue the United States and Australia four years ago in the ICJ, but held off for unspecified reasons.[177] In the corporate world, use of the shareholder proposal process and SEC disclosure requirements have been suggested as ways of forcing the issue.[178]

New categories of litigation also may emerge. For example, the head of the California Air Resources Board has predicted a court challenge to her state's cap-and-trade system to regulate GHGs (expected to take effect in 2012). Such a challenge, she said, might argue that the cap-andtrade system's fee on GHG emissions imposes a new tax, which requires a 2/3 vote of the state legislature. As another example, rising sea levels may prompt lawsuits seeking a judicial blessing for the landward migration of the public's beach access rights.[179] And of course, any climate change legislation enacted by Congress is likely to spawn its own generation of litigation.[180]

Whether these new paths will yield results, only time will tell. It is clear, however, that if there is to be a government response to climate change at all, a solution from the political branches is more likely to be comprehensive and fully reflective of societal priorities than the typically narrowly targeted results of litigation. Many proponents of litigation or unilateral action by the states freely concede that such initiatives are make-do efforts that, while making a small contribution to mitigating climate change, are also aimed at prodding the national government to act.

End Notes

[1] Robert Cook, Obama Said to Be "Off to Fast Start" With Economic Stimulus Legislation, BNA Daily Env't Rept. (March 18, 2009).

[2] 549 U.S. 497 (2007).

[3] Similar ground is covered by Justin R. Pidot, *Global Warming in the Courts: An Overview of Current Litigation and Common Legal Issues* (Georgetown Environmental Law and Policy Institute 2006) (available, together with a March, 2007 update, at http://www.law.georgetown.edu/gelpi/), and Todd O. Madden and Eric McLaughlin, *Climate Change Litigation: Trends and Developments*, BNA Daily Env't Rpt. B-1 (April 3, 2007). A regularly updated chart of climate change cases, prepared by Michael Gerrard, Director, Center for Climate Change Law, Columbia University, is available at http://www.climatecasechart.com. A useful blog is the Constitutional Accountability Center's Warming Law: Changing the Climate in the Courts, found at http://theusconstitution.org/blog.warming/. Broader treatments of the legal implications of climate change may be found in Michael Gerrard (ed.), GLOBAL CLIMATE CHANGE AND U.S. LAW (ABA 2007), and at least three law-review symposium issues: *Responses to Global Warming: The Law, Economics, and Science of Climate Change*, 155 U. Pa. L. Rev. 1353 (2007); *Changing Climates: Adapting Law and Policy to a Transforming World*, 55 UCLA L. Rev. 1479 (2008); and *Federalism and Climate Change: The Role of the States in a Future Federal Regime*, 50 Ariz. L. Rev. 673 (2008).

[4] An example of a case that deals with climate change only implicitly (at least so far) is *State of New York v. U.S. Dep't of Energy*, No. 08-0311 (2d Cir. filed January 17, 2008), in which three states (NY, CN, MA) challenge the Department's energy conservation standards for residential furnaces and boilers. Though we are given to understand that the climate change benefits of reducing fossil fuel consumption by such furnaces and boilers was a

consideration in filing suit, the petition for review does not mention CO2 or climate change, and thus we do not include this case in the body of the report.

[5] Kyoto Protocol to the United Nations Framework Convention on Climate Change, concluded December 10, 1997, U.N. Doc. FCC/CP/1997/L.7 Add. 1, reprinted at 37 I.L.M. 22 (1998). One indication of Senate antipathy to the Kyoto Protocol was its adoption by 95-0 of the so-called Byrd-Hagel resolution urging the President not to sign any international agreement on climate change that would result in serious injury to the U.S. economy or that did not include provisions regarding the GHG emissions of developing countries. S.Res. 98, 105th Congress (1997).

[6] *See generally* CRS Report 98-349, *Global Climate Change: Selected Legal Questions About the Kyoto Protocol*, by David M. Ackerman. This report concluded that "there does not appear to be any clear legal authority that could be invoked to sustain the provisional application of the Kyoto Protocol." *Id.* at 6.

[7] P.L. 105-276, 112 Stat. at 2496 (1998) (barring EPA's use of FY1999 funds to implement Protocol); P.L. 106-74, 113 Stat. at 1080 (1999) (same for FY2000); P.L. 106-377, 114 Stat. at 1141A-41 (2000) (same for FY2001).

[8] Memorandum from Jonathan Z. Cannon, EPA General Counsel, to Carol M. Browner, EPA Administrator, EPA's Authority to Regulate Pollutants Emitted by Electric Power Generation Sources (April 10, 1998).

[9] *Is CO2 A Pollutant and Does EPA Have the Power to Regulate It?*, Joint Hearing Before the Subcomm. on National Environmental Growth, Natural Resources and Regulatory Affairs of the House Comm. on Gov't Reform and the Subcomm. on Energy and Environment of the House Comm. on Science, 106th Cong. (1999).

[10] Testimony of Gary Guzy, Joint Hearing, *supra* note 9, at 11.

[11] *See* Veronique Bugnion and David M. Reiner, *A Game of Climate Chicken: Can EPA Regulate Greenhouse Gases Before the U.S. Senate Ratifies the Kyoto Protocol?*, 30 Envtl. L. 491 (2000).

[12] Civ. Action No. 3:03CV984 (PCD) (D. Conn.) (filed June 4, 2003).

[13] 42 U.S.C. § 7408.

[14] 42 U.S.C. § 7409.

[15] *See* CAA § 111, 42 U.S.C. § 7411.

[16] 71 Fed. Reg. 9,866 (February 27, 2006).

[17] Coke Oven Environmental Task Force v. EPA, No. 06-1131 (D.C. Cir. filed April 7, 2006).

[18] No. 06-1322.

[19] NESHAPSs are governed by CAA section 112, 42 U.S.C. § 7412.

[20] CAA § 111(b)(1)(B); 42 U.S.C. § 7411(b)(1)(B).

[21] 73 Fed. Reg. 35,838 (June 24, 2008), codified at 40 C.F.R. Part 60, Subpart Ja. The agency's response to commenters wanting EPA to promulgate an NSPS for GHGs as part of its rule revisions is at 35858-35860. The agency's response is interesting because owing to the Supreme Court decision in *Massachusetts v. EPA* the previous year, EPA could no longer argue that the term "air pollutant" in section 111 does not reach GHGs. One argument used by EPA is that the eight-year reviews of NSPSs required by CAA section 111(b)(1)(B) do not mandate promulgation of NSPSs for pollutants not already covered by the NSPS under review. EPA conceded that it had promulgated NSPSs for previously uncovered pollutants in the past, but argued that this was discretionary. It is better, the agency asserted, to address GHG emissions through the process begun by its Advance Notice of Proposed Rulemaking.

[22] *See* CAA § 307(d)(7)(B), 42 U.S.C. § 7607(d)(7)B) (when EPA Administrator may convene proceeding for reconsideration of rule). One petition, from the Environmental Integrity

[23] *Id.*
[24] 434 F. Supp. 2d 957 (D. Or. 2006).
[25] CAA § 165, 42 U.S.C. § 7475.
[26] PSD areas are areas that are either attaining the National Ambient Air Quality Standard for the pollutant in question or are unclassifiable for that pollutant. CAA § 161, 42 U.S.C. § 7471. The PSD portion of the CAA, 42 U.S.C. §§ 7470-7492, sets limits on the degree to which ambient concentrations of a pollutant will be allowed to rise toward the cap set by the National Ambient Air Quality Standard for that pollutant.
[27] "Major emitting facility" is defined at CAA section 169(1), 42 U.S.C. § 7479(1). "Best available control technology" is defined at CAA section 169(3), 42 U.S.C. § 7479(3).
[28] CAA § 165(a)(4), 42 U.S.C. § 7475(a)(4); CAA § 169(3), 42 U.S.C. § 7479(3).
[29] 40 C.F.R. Part 75.
[30] 40 C.F.R. § 52.21(b)(50) (defining "Regulated NSR pollutant").
[31] The boiler would have functioned as a cogeneration unit providing both electrical power and heat to the university by burning wood, coal, and natural gas.
[32] 42 U.S.C. § 7521(a)(1).
[33] EPA, Control of Emissions from New Highway Vehicles and Engines, 68 Fed. Reg. 52,922 (September 8, 2003). EPA's ruling followed a suit by the original petitioners alleging unreasonable delay. Center for Technology Assessment v. Whitman, No. 02-CV-2376 (D.D.C. filed December 5, 2002).
[34] Memorandum from Robert E. Fabricant, EPA General Counsel, to Marianne L. Horinko, EPA Acting Administrator, EPA's Authority to Impose Mandatory Controls to Address Global Climate Change Under the Clean Air Act (August 28, 2003).
[35] 415 F.3d 50 (D.C. Cir. 2005).
[36] 549 U.S. 497 (2007).
[37] *Id.* at 504-505.
[38] CAA § 307(b)(1), 42 U.S.C. § 7607(b)(1).
[39] 549 U.S. at 517-18.
[40] *Id.* at 518.
[41] *Id.*
[42] *Id.* at 524.
[43] *Id.* at 525.
[44] 49 U.S.C. § 32902.
[45] 549 U.S. at 501.
[46] Three weeks after the decision in *Massachusetts v. EPA*, the Senate held a hearing devoted exclusively to it: *The Implications of the Supreme Court's Decision Regarding EPA's Authorities with Respect to Greenhouse Gases Under the Clean Air Act*, Hearing Before the Senate Comm. on Env't and Pub. Works (April 24, 2007) (hereinafter Senate hearing).
[47] Justice Scalia's dissent characterizes EPA's three options similarly: 127 S. Ct. at 1472.
[48] See CRS Report RL34266, *Climate Change: Science Highlights*, by Jane A. Leggett.
[49] *Senate hearing, supra* note 46 (prepared statement of EPA Administrator Stephen Johnson). The EPA Administrator was apparently referring to the Court's statement that "We need not and do not reach the question ... whether policy concerns can inform EPA's actions in the event that it makes [an endangerment finding]." 549 U.S. at 534-535.
[50] *Briefing by Conference Call on the President's Announcement on CAFE and Alternative Fuel Standards*, May 14, 2007 (statement of EPA Administrator Stephen Johnson), available at

whitehouse.gov/news/releases/2007. *See also* Exec. Order No. 13432, *Cooperation Among Agencies in Protecting the Environment with Respect to Greenhouse Gas Emissions from Motor Vehicles, Nonroad Vehicles, and Nonroad Engines*, 72 Fed. Reg. 27,717 (May 16, 2007).

[51] P.L. 110-140.

[52] Massachusetts v. EPA, No. 03-1361 (D.C. Cir. June 26, 2008). The court provided no explanation of its decision, except for an opinion by Judge Tatel concurring in part and dissenting in part. Judge Tatel agreed with the other two judges that no writ of mandamus was yet justified. Still, he would have held the petitioners' motion in abeyance and required periodic updates from the agency because its postponement was indefinite.

[53] 73 Fed. Reg. 44,354, 44,355 (July 30, 2008).

[54] *Id.*

[55] *See, e.g.*, CAA § 108(a)(1)-(2), 42 U.S.C. § 7408(a)(1)-(2) (requiring the EPA Administrator to maintain a list of each "air pollutant, emissions of which, in his judgment, cause or contribute to air pollution which may reasonably be anticipated to endanger public health or welfare," and then issue air quality criteria and national ambient air quality standards for that air pollutant).

[56] CAA § 165, 42 U.S.C. § 7475.

[57] CAA § 165(a)(4), 42 U.S.C. § 7475(a)(4).

[58] 73 Fed. Reg. at 44,399.

[59] *See* Sara Zdeb, *From* Georgia v. Tennessee Copper *to* Massachusetts v. EPA: Parens Patriae Standing for State Global-Warming Plaintiffs, 96 Geo. L. J. 1059 (2008).

[60] California v. General Motors Corp., 2007 Westlaw 2726871 (N.D. Cal. September 17, 2007). This case is discussed in Section V.

[61] 16 U.S.C. §§ 1361-1421h.

[62] 16 U.S.C. § 1371(a)(5). In the MMPA, "take" means "to harass, hunt, capture, or kill" any marine mammal, or attempt to do so. 16 U.S.C. § 1362(13).

[63] 71 Fed. Reg. 43,926 (August 2, 2006).

[64] In a case of the same name, *Center for Biological Diversity v. Kempthorne*, No. 07-5109 (N.D. Cal. filed October 4, 2007), environmental groups challenge the Secretary of the Interior's failure to issue updated stock assessment reports for marine mammals under his jurisdiction (sea otters, polar bears, walrus, and manatees) within the time frames mandated by the MMPA. The complaint asserts as examples that since the last stock assessment reports on the polar bear and walrus, "global warming has caused the loss of sea ice upon which [those species] depend...." The case was settled in late 2008, with deadlines for new stock assessments.

[65] 16 U.S.C. §§ 1531-1544. The ESA defines "take" similarly to the MMPA, *see supra* note 62. 16 U.S.C. § 1532(19).

[66] *See* John Kostyak and Dan Rohlf, Conserving Endangered Species in an Era of Global Warming, 38 Envtl. L. Rptr. 10203 (Apr. 2008); Sarah J. Morath, The Endangered Species Act: A New Avenue for Climate Change Litigation?, 29 Pub. Land & Res. L. Rev. 23 (2008); Ari Sommer, Student Note, Taking the Pit Bull Off the Leash: Siccing the Endangered Species Act on Climate Change, 36 B.C. Envtl. Affairs L. Rev. 273 (2009).

[67] 16 U.S.C. § 1538(a)(1)(B)-(C).

[68] By general rule, the Fish and Wildlife Service has extended all of the endangered species prohibitions to threatened animals. 50 C.F.R. § 17.31. "Special rules," withdrawing particular threatened species from aspects of the general regime, have been promulgated for those species with atypical management needs, such as grizzly bears. 50 C.F.R. § 17.40(b).

[69] 16 U.S.C. § 1536(a)(2). Because section 7 is more easily triggered when the species' habitat receives a formal designation as "critical habitat," litigation to compel such designation is another aspect of environmental groups' efforts to use the ESA against global warming. *See* ESA § 4(a)(3), 16 U.S.C. § 1533(a)(3).

[70] 89 Fed. Reg. 26,852 (May 9, 2006). The Center for Biological Diversity has also settled a suit requiring NMFS to designate critical habitat for ESA-listed corals. The final critical habitat rule is at 73 Fed. Reg. 72,209 (Nov. 26, 2008).

[71] 73 Fed. Reg. 28,211 (May 15, 2008), codified at 50 C.F.R. § 17.11(h).

[72] Center for Biological Diversity v. Kempthorne, No. 08-1339 (N.D. Cal. April 28, 2008).

[73] *See* CRS Report RL33941, *Polar Bears: Listing Under the Endangered Species Act*, by Eugene H. Buck, M. Lynne Corn, and Kristina Alexander.

[74] 74 Fed. Reg. 1,937 (January 14, 2009).

[75] Listing of the ribbon seal was denied by NMFS on December 30, 2008 (73 Fed. Reg. 79,822). CBD has filed its 60- day notice of intent to sue.

[76] *See generally* Brendan R. Cummings and Kassie R. Siegel, Ursus maritimus: *Polar Bears on Thin Ice*, Natural Res. & Env't (ABA) 3 (Fall 2007) (discussing how "the listing process for the polar bear highlights the possibilities and limitations of using the ESA to address otherwise unregulated GHG emissions").

[77] 50 C.F.R. §§ 17.3 (Fish and Wildlife Service), 222.102 (NOAA Fisheries).

[78] 73 Fed. Reg. 28,306 (May 15, 2008) (interim final rule); 73 Fed. Reg. 76,249 (Dec. 16, 2008) (final rule). Codified at 50 C.F.R. § 17.40(q)(4). Special rules, also known as "4(d) rules," are authorized by ESA section 4(d) for threatened (not endangered) species that are considered to have special management needs. (By regulation, other threatened species receive the same protections that endangered species do.) The ESA permits considerable flexibility in the crafting of 4(d) rules, demanding only that they be "necessary and advisable to provide for the conservation of [the threatened] species." 16 U.S.C. § 1533(d).

[79] In re Polar Bear Endangered Species Act Listing and 4(d) Rule Litigation, No. 1:08-mc-764. For background, see CRS Report RL34573, *Does the Endangered Species Act (ESA) Listing Provide More Protection of the Polar Bear?: A Look at the Special Rules*, by Kristina Alexander.

[80] 73 Fed. Reg. 76,272 (Dec. 16, 2008). Codified at 50 C.F.R. § 402.03(b).

[81] Codified at 50 C.F.R. § 402.02.

[82] Center for Biological Diversity v. Kempthorne, No. C 08-05546 MHP; Natural Resources Defense Council v. U.S. Department of the Interior, C 08-05605 MMC; National Wildlife Federation v. Kempthorne, No. C 08-05654 SI.

[83] People of the State of California v. Kempthorne, No. C 08-05775 EMC.

[84] Congress has entered the fray as well. H.R. 1431, sec. 306(b), amends the ESA by adding a new sentence: "The impact of greenhouse gas on any species of fish or wildlife or plant shall not be considered for any purpose in the implementation of this Act."

[85] Omnibus Appropriations Act for FY 2009, P.L. 111-8, Div. E, § 429.

[86] 49 U.S.C. § 32902(a).

[87] 538 F.3d at 1199.

[88] Center for Biological Diversity v. National Highway Traffic Safety Admin., No. 09-70972 (9th Cir. filed April 3, 2009).

[89] 43 U.S.C. § 1331 et seq.

[90] Taken from Petitioner Center for Biological Diversity's Non-Binding Statement of Issues, filed August 3, 2007.

[91] An apparently exhaustive survey of the NEPA/climate change cases, decided and pending, is Joseph Mendelson III (Legal Director, Center for Food Safety and International Center for Technology Assessment), *Surveying the National Environmental Policy Act and the Emerging Issues of Climate Change, Genetic Engineering and Nanotechnology* (October 30, 2007) (copy on file with author).

[92] *See also* International Center for Technology Assessment et al., *Petition requesting that the Council on Environmental Quality amend its regulations to clarify that climate change analyses be included in environmental review documents* (filed February 28, 2008). The Center for American Progress argues that President Obama should issue an executive order instructing federal agencies to consider climate change in their NEPA-mandated documents. Nancy Sutley, the chairwoman of CEQ appointed by President Obama, reportedly has said she will be considering the issue in response to both informal requests from federal agencies and the International Center for Technology Assessment petition. *See generally* Conor O'Brien, Student Note, *I Wish They All Could Be California Environmental Quality Acts: Rethinking NEPA in Light of Climate Change*, 36 B.C. Envtl. Affairs L. Rev. 239 (2009).

[93] Other model years were involved, too, but only the challenge to the model year 1989 CAFE standard involved a climate change argument.

[94] 912 F.2d at 501.

[95] Florida Audubon Society v. Bentsen, 94 F.3d 658 (D.C. Cir. 1996). The four dissenting judges argued that the majority had "misapplied the doctrine of standing to the assertion of a procedural right, such as the preparation of an EIS, with the consequence that it will be effectively impossible for anyone to bring a NEPA claim in the context of a rulemaking with diffuse impact." *Id.* at 673.

[96] *See, e.g.*, Friends of the Earth v. Laidlaw Environmental Services, 528 U.S. 167 (2000).

[97] 794 F.2d at 401.

[98] In finding standing, the judge repudiated an earlier climate change/standing decision of the same court. In *Center for Biological Diversity v. Abraham*, 218 F. Supp. 2d 1143 (N.D. Cal. 2002), plaintiffs had sought enforcement of the Energy Policy Act as it related to the acquisition of alternative fuel vehicles by the United States. In rejecting standing, this decision spurned plaintiffs' climate change concerns as "too general, too unsubstantiated, too unlikely to be caused by defendant's conduct, and/or too unlikely to be redressed by the relief sought to confer standing." In *Friends of the Earth*, the court neutralized this pronouncement by noting that "*Center for Biological Diversity* was decided before the Ninth Circuit clarified in [Citizens for Better Forestry v. U.S. Dep't of Agriculture, 341 F.3d 961, 972 (9th Cir. 2003)] that environmental plaintiffs raising procedural concerns need not present proof that the challenged federal project will have particular environmental effects."

[99] By the time of settlement, the case was styled *Friends of the Earth v. Spinelli*.

[100] An organization has standing to bring suit on behalf of its members when "(a) its members would otherwise have standing to sue in their own right; (b) the interests it seeks to protect are germane to the organization's purpose; and (c) neither the claim asserted nor the relief requested requires the participation of individual members in the lawsuit." Hunt v. Washington State Apple Advertising Comm'n, 432 U.S. 333, 343 (1977).

[101] 260 F. Supp. 2d at 1028.

[102] 71 Fed. Reg. 17,566 (April 6, 2006).

[103] 508 F.3d at 550.

[104] *Id.*

[105] 71 Fed. Reg. 43,926 (August 2, 2006).

[106] 40 C.F.R. § 1508.8.
[107] *See* 345 F.3d at 550.
[108] *See* Dave Owen, Climate Change and Environmental Assessment Law, 33 Colum. J. Env'l L. 57 (2008).
[109] The attorney general lawsuit is State of California v. County of San Bernardino, No. CIV-SS07-00329 (Cal. Super. Ct. filed April 12, 2007).
[110] Confidential Settlement Agreement, available at http://ag.ca.gov/cms_pdfs/press/2007-08-21_San_Bernardino_settlement_agreement.pdf.
[111] 15 U.S.C. §§ 2921-2961.
[112] *See* 15 U.S.C. § 2931(a) (congressional findings suggestive of the act's scope).
[113] 15 U.S.C. § 2934(a).
[114] 15 U.S.C. § 2936.
[115] The summary was published at 72 Fed. Reg. 73,771 (December 28, 2007).
[116] 5 U.S.C. § 552.
[117] Renamed the House Committee on Oversight and Government Reform early in the 110th Congress.
[118] *Political Interference with Climate Change Science Under the Bush Administration*, available at http://oversight
[119] On file with author.
[120] This nutshell on possible plaintiffs and defendants is adapted from David Hunter and James Salzman, *Negligence in the Air: The Duty of Care in Climate Change Litigation*, 155 U. Pa. L. Rev. 1741, 1750-1752 (2007).
[121] *See generally* Thomas W. Merrill, *Global Warming as a Public Nuisance*, 30 Colum. J. Envtl. L. 293 (2005); Matthew F. Pawa, *Global Warming: The Ultimate Public Nuisance*, 39 Envtl. Law Rptr. 10230 (March 2009); Jim Gitzlaff, *Getting Back to Basics: Why Nuisance Claims Are of Limited Value in Shifting the Costs of Climate Change*, 39 Envtl. Law Rptr. 10218 (March 2009).
[122] American Electric Power Co., Inc., The Southern Co., Cinergy Corp., Tennessee Valley Authority, and Xcel Energy, Inc.
[123] An activity is a public nuisance if it creates an unreasonable interference with a right common to the general public. Unreasonableness may rest on the fact that the activity significantly interferes with public health and safety, or has produced a permanent or long-lasting effect and, as the actor knows or has reason to know, has a significant effect on the public right. RESTATEMENT (SECOND) OF TORTS § 821B (1979).
[124] An activity is a private nuisance if it is a nontrespassory invasion of another's interest in the private use and enjoyment of land. *Id.* at § 821D.
[125] *See* Vincent S. Oleskiewicz and Douglas B. Sanders, *The Advent of Climate Change Litigation Against Corporate Defendants*, BNA Daily Env't Rpt. B-1 (November 15, 2004). The authors review the *State of Connecticut* and *Open Space Institute* suits in some detail, assess the defenses available in tort-based climate change suits generally, and extract clues as to the potential success of such litigation from the history of litigation against tobacco companies.
[126] An interesting question raised by the Prof. Merrill article, *supra* note 121, is whether these general standing requirements, developed in the context of private lawsuits, should apply in a suit such as *State of Connecticut*—that is, in a parens patriae suit brought by state attorneys general under public nuisance law.
[127] Connecticut v. American Elec. Power, 406 F. Supp. 2d 265 (S.D.N.Y. 2005).
[128] Baker v. Carr, 369 U.S. 186, 210 (1962).

[129] *Id.* at 217.

[130] Presumably two of the plaintiffs, New York State and New York City, have been able to support their standing to sue by arguments not contrary to those they made *against* plaintiff standing in another public-nuisance climate change case in which they were the defendants. Avoiding contradictory arguments was presumably facilitated by the idiosyncratic nature of the global-warming harms alleged by the *pro se* plaintiff—e.g., those based on plaintiff's enhanced vulnerability to disease-causing pollution as compared to the general population. According to the court, plaintiff appeared to be requesting an injunction ordering the defendants to stop polluting and use his invention for reducing carbon dioxide emissions. Korsinsky v. U.S. EPA, 192 Fed. Appx. 171 (2d Cir. 2006) (affirming district court dismissal based on lack of standing).

[131] CAA § 116, 42 U.S.C. § 7416. The exceptions in this nonpreemption provision say that states may not adopt emission limitations for stationary sources that are *less* stringent than those in state implementation plans, new source performance standards, or national emission standards for hazardous air pollutants.

[132] *See* Global Warming Solutions Act of 2006, A.B. 32, Cal. Health & Safety Code § 38500 et seq. This law requires that GHG emission limits be in effect in California by 2012 to reduce statewide GHG emissions to the 1990 level by 2020. Note, however, that although A.B. 32 applies chiefly to stationary sources, it provides that if the mobile source GHG emission limits imposed by an earlier state enactment are struck down, "alternative regulations" to restrict mobile-source GHG emissions shall be imposed under A.B. 32. As the following paragraphs of the text discuss, this earlier enactment is now the subject of a preemption challenge.

[133] CAA § 209(a), 42 U.S.C. § 7543(a).

[134] CAA § 302(k), 42 U.S.C. § 7602(k). Emphasis added.

[135] CAA § 209(b), 42 U.S.C. § 7543(b). Under section 209(b), EPA "shall" grant the waiver "if the State determines that the State standards will be, in the aggregate, at least as protective of public health and welfare as applicable federal standards." However, no waiver shall be granted if EPA finds that the state's determination is arbitrary and capricious, the state does not need the standards to meet "compelling and extraordinary conditions," or the state standards and accompanying enforcement procedures are inconsistent with CAA section 202(a).

[136] CAA § 177, 42 U.S.C. § 7507. Section 177 limits its extension of the section 209 waiver to those states that have approved nonattainment-area plans. This includes all states except North Dakota, South Dakota, and Hawaii.

[137] Cal. Heath & Safety Code § 43018.5.

[138] 73 Fed. Reg. 12,156 (March 6, 2008).

[139] *See generally* CRS Report RL34099, *California's Waiver Request Under the Clean Air Act to Control Greenhouse Gases From Motor Vehicles*, by James E. McCarthy and Robert Meltz.

[140] 74 Fed. Reg. 7,040 (February 12, 2009).

[141] *See generally* Kristien G. Knapp, *The Legality of EPA's Greenhouse Gas Waiver Denial*, 39 Envtl. L. Rptr. 10127 (February 2009).

[142] EPCA's fuel economy provisions are at 49 U.S.C. §§ 32901-32919.

[143] *Id.* at § 32902(b)(2)(A).

[144] *Id.* at § 32919.

[145] *Id.* at § 32902(f).

[146] 549 U.S. at 532.

[147] The court made clear its discomfort in second-guessing the AUJ on a scientific matter unfamiliar to the court.
[148] See supra note 26.
[149] Kansas Stat. Ann. § 65-3012.
[150] The "Dormant Commerce Clause" is a judicially created corollary of the Commerce Clause in Art. I, section 8 of the U.S. Constitution. It asserts that states may not impose undue burdens on interstate commerce.
[151] See, e.g., Sean B. Hecht, *Climate Change and the Transformation of Risk,: Insurance Matters*, 55 UCLA L. Rev. 1559 (2008); Justin Pidot, Georgetown Environmental Law and Policy Inst., COASTAL DISASTER INSURANCE IN THE ERA OF GLOBAL WARMING,: THE CASE FOR RELYING ON THE PRIVATE MARKET (2007).
[152] Andrew L. Strauss, *The Legal Option: Suing the United States in International Forums for Global Warming Emissions*, 33 Envtl. L. Rptr. 10185 (2003).
[153] Agreement for the Implementation of the Provisions of the United Nations Convention on the Law of the Sea of 10 December 1982 Relating to the Conservation and Management of Straddling Fish Stocks and Highly Migratory Fish Stocks, 34 Int'l Legal Materials 1547. The United States is a party to this agreement and, by reference to the Law of the Sea Convention, it incorporates binding dispute-resolution mechanisms.
[154] RESTATEMENT (THIRD) OF FOREIGN RELATIONS LAW § 601(1). *See also* Legality of the Threat or Use of Nuclear Weapons, Advisory Opinion, 1996 ICJ Reports 226, 241-242 (July 8, 1996) ("the existence of the general obligation of states to ensure that activities within their jurisdiction and control respect the environment of other states or of areas beyond national control is now part of the corpus of international law relating to the environment").
[155] Trail Smelter (U.S. v. Canada), 3 R.I.A.A. 1938, 1965 (March 11, 1941).
[156] For an eight-page summary of the 176-page petition, go to http://earthjustice.org/library/legal_docs/summary-ofinuit-petition-to-inter-american-council-on-human-rights.pdf.
[157] See generally Sara C. Aminzadeh, Note, *A Moral Imperative: The Human Rights Implications of Climate Change*, 30 Hastings Int'l & Comp. L. Rev. 231 (2007).
[158] Convention Concerning the Protection of the World Cultural and Natural Heritage, art. 8, signed November 23, 1972, entered into force December 17, 1975, 27 U.S.T. 37.
[159] *Id.* at art.11, par. 4.
[160] *Id.* at art. 4.
[161] World Heritage Comm. Decision 30 COM 7.1, adopted July 10, 2006, available at http://law.lclark.edu/org/ ielp/glacierpetition.html.
[162] Filed January 29, 2009, by Earthjustice and the Australian Climate Change Program.
[163] Petition at 1.
[164] Further, the petition requests that advisory bodies to the World Heritage Convention, State Parties, and site managers undertake studies on the sources of black carbon that are polluting high latitude and high altitude sites and recommend measures to reduce such emissions. It then requests the World Heritage Committee to develop a program of corrective measures.
[165] *See, e.g.*, J. Macabrey, Researchers Warn That Sea Levels Will Rise Much Faster Than Expected, ClimateWire (E&E Publishing March 11, 2009), available at eenews.net/climatewire/2009/03/11/1.
[166] Massachusetts v. EPA, 549 U.S. 497 (2007) (see Section I of this report).
[167] Green Mountain Chrysler Plymouth Dodge Jeep v. Crombie, 508 F. Supp. 2d 295 (D. Vt. 2007), and Central Valley Chrysler-Jeep, Inc. v. Goldstene, 529 F. Supp. 2d 1151 (E.D. Cal.

2007), both described in section VI of this report; Center for Biological Diversity v. National Highway Traffic Safety Administration, 508 F.3d 508 (9th Cir. 2007), described in sections III and IV of this report.

168 *Green Mountain, supra* note 167.

169 *Center for Biological Diversity, supra* note 167.

170 Myles R. Allen and Richard Lord, *The Blame Game: Who Will Pay for the Damaging Consequences of Climate Change?*, 432 Nature 551 (December 2004); David A Grossman, *Warming Up to a Not-So-Radical Idea: Tort-Based Climate Change Litigation*, 28 Colum. J. Envtl. L. 1 (2003): Eduardo M. Penalver, *Acts of God or Toxic Torts? Applying Tort Principles to the Problem of Climate Change*, 38 Nat. Res. J. 563, 569 (1998).

171 *See generally* Richard J. Lazarus, THE MAKING OF ENVIRONMENTAL LAW ch. 1 (2004).

172 *See, e.g.*, Kristin Marburg, *Combating the Impacts of Global Warming: A Novel Legal Strategy*, 2001 Colo. J. Int'l L. & Pol'y 171 (2001).

173 *See, e.g.*, California's A.B. 32, the Global Warming Solutions Act of 2006, *supra* note 132. *See generally* CRS Report RL33812, *Climate Change: Action by States to Address Greenhouse Gas Emissions*, by Jonathan L. Ramseur.

174 For example, RGGI is an initiative involving 10 northeastern states to stabilize CO2 emissions from power plants at 188 million tons per year from 2009-2014 and then to reduce emissions by 2.5% per year over the next four years.

175 Congressional Green Sheets Newsroom, December 17, 2004. The same source reports that Rep. Joe Barton (R-Texas), then-chairman of the House Energy and Commerce Committee, said that any international compact involving state governments would have to be approved by Congress and that "we would tend to look at it with a lot of skepticism." Kenneth Colburn, who is helping to coordinate the states' effort, was said to question the need for federal authorization, on the theory that any transatlantic trades would be commercial transactions, not government-togovernment.

176 United Kingdom and California Announcement on Climate Change and Clean Energy Collaboration.

177 *See* http://www.tuvaluislands.com. Tuvalu alleged that Australia is the biggest per capita producer of GHGs, and that the United States is the biggest single emitter. *See also* Aurelie Lopez, *The Protection of Environmentally Displaced Persons in International Law*, 37 Envtl. L. 365, 372-373 (2007). Residents of the Alaskan village of Shishmaref on the Bering Strait, who are now being relocated, are apparently the first American climate change refugees.

178 *See, e.g.*, Sung Ho (Danny) Choi, Note, *It's Getting Hot in Here: The SEC's Regulation of Climate Change Shareholder Proposals Under the Ordinary Business Exception*, 17 Duke Envtl. L. & Pol'y Forum 165 (2006); California Public Employees' Retirement System et al., *Petition for interpretive guidance on climate change disclosure*, SEC No. 4-547 (submitted September 18, 2007); Free Enterprise Action Fund, *Petition for interpretive guidance under the Securities Act of 1933 that would require registrants to disclose to shareholders the business risks of laws and regulations intended to address global warming concerns*, SEC No. 4-549 (submitted October 22, 2007).

179 *See, e.g.*, Severance v. Patterson, 485 F. Supp. 2d 793, 804 (S.D. Tex. 2007) (finding no property rights taking based on state's migrating easement allowing public access to the dry beach between mean high tide line and natural vegetation line, notwithstanding that these lines move).

[180] The just-released discussion draft of the Waxman-Markey energy/climate-change bill, titled the American Clean Energy and Security Act of 2009, expands the existing citizen suit provision in the Clean Air Act to facilitate suits based on climate change (currently, draft bill section 336). The amendments are geared toward lowering the barriers to standing often encountered by climate change plaintiffs—the barriers that were lowered in *Massachusetts v. EPA* in the special circumstance where the citizen plaintiff is a state. Thus, the draft provision states that persons entitled to file citizen suits include those who suffer, or reasonably expect to suffer, "the incremental exacerbation of any effect or risk that is associated with a small incremental emission of any air pollutant (including any greenhouse gas ...), whether or not the effect or risk is widely shared."

INDEX

A

ABA, 38, 39, 117, 121
abatement, 15, 53
Abraham, 122
access, viii, 2, 22, 31, 42, 117, 126
accounting, 11
acid, 83, 116
adaptation, viii, 2, 5, 25, 112
adaptations, 41
Administrative Procedure Act, 97, 102
adverse effects, 112
affirming, 34, 124
agencies, 11, 12, 19, 35, 36, 39, 65, 68, 71, 75, 98, 99, 100, 102, 122
agency actions, x, 64, 65, 93, 97
agriculture, 19
air pollutants, x, 17, 63, 64, 84, 87, 107
air quality, 36, 75, 81, 120
Alaska, 9, 36, 39, 55, 58, 90, 93, 99, 105
ambient air, 8, 81, 89, 120
amortization, 31, 42
appointees, 102
Appropriations Act, 35, 121
arbitration, 17, 113
articulation, 49
Asia, 115
assessment, 36, 98, 99, 100, 101, 120
asylum, 33, 43
atmosphere, viii, x, 2, 11, 13, 36, 46, 57, 58, 67, 86
automobiles, 37, 68, 108
aversion, ix, 46
avoidance, 29
awareness, 42

B

barriers, vii, 1, 3, 96, 98, 127
benefits, xi, 28, 78, 94, 117
bias, 69
bioenergy, 72
biomass, 72
blogs, 38
boilers, 117
building code, 41
businesses, 16
buyers, 23

C

candidates, 3
Capitol Hill, 46
carbon, vii, xi, 2, 10, 11, 37, 53, 59, 78, 80, 86, 94, 96, 98, 114, 124, 125
carbon emissions, xi, 78, 94
case law, 9, 24, 27, 30, 32
category a, 81
category b, 79

cattle, 100
causal relationship, 86
causation, 3, 47, 56, 86, 97, 104, 106, 116
CBD, 92, 94, 95, 121
certification, 41
CGL, 16
challenges, xi, 7, 9, 10, 13, 27, 29, 78, 84, 86, 95, 116
chemical, ix, 15, 46, 56, 106
chemicals, 31, 116
Chief Justice, 87
children, 57
China, 37
citizens, 34, 103
City, 36, 37, 38, 39, 40, 41, 42, 52, 96, 98, 103, 110, 124
classes, 107
clean air, 90
Clean Air Act, vii, ix, x, xi, 1, 3, 5, 14, 17, 34, 35, 45, 46, 47, 53, 58, 63, 64, 66, 70, 74, 75, 76, 77, 78, 79, 80, 84, 89, 115, 119, 124, 127
Clean Water Act, 32, 35
climate change issues, 78, 79, 116
Clinton Administration, 80, 85
CO_2, 11, 13, 15, 52, 57, 69, 72, 74, 80, 81, 82, 83, 84, 85, 89, 98, 99, 103, 107, 110, 111, 118, 119, 126
coal, xi, 10, 13, 15, 56, 78, 79, 83, 84, 97, 100, 106, 111, 119
coastal management, 41
cogeneration, 119
collaboration, 87
collateral, 109
combustion, 72, 103
commerce, 12, 13, 37, 125
commercial, 90, 102, 111, 126
community, 28, 30
compensation, 19, 22, 28, 36, 47
competitors, 13
complexity, 54, 89, 105
compliance, 54, 69, 100
compulsion, 8
conditioning, 73

conflict, 19, 33, 109
consensus, 3, 14, 16, 24, 25, 29
consent, 27
conservation, 19, 101, 113, 121
Consolidated Appropriations Act, 67
conspiracy, 55, 56, 105, 106
Constitution, 3, 13, 18, 50, 51, 125
constitutional issues, 22
construction, 23, 29, 30, 32, 35, 72, 75, 83, 84, 89, 93, 97, 111
consumption, 19, 100
contamination, 116
controversies, 3
cooperation, 35, 114
corporate average fuel economy, xi, 12, 78, 94, 96, 99, 108
cost, 55, 94, 110
cost-benefit analysis, 94
Court of Appeals, 65
covering, 69, 84
crop, viii, 2, 3
culture, 18, 113

D

danger, 102, 114
decision control, 15
decision makers, 11
decomposition, 72
defendants, 4, 15, 16, 47, 49, 50, 52, 53, 54, 55, 98, 101, 102, 103, 104, 105, 106, 115, 116, 123, 124
degradation, 35, 93
Delta, 91
denial, xi, 5, 26, 31, 64, 66, 67, 78, 81, 85, 87, 108, 111
Department of Agriculture, 97, 100
Department of Commerce, 35
Department of the Interior, 35
Department of Transportation, 32, 87
deposits, 20
destruction, 35, 91
devaluation, 24
developing countries, 87, 112, 118
developing nations, 88

directives, 104
disaster, 28, 30, 32
disclosure, 30, 68, 116, 126
discomfort, 125
discrimination, 13
displacement, ix, 5, 16, 33, 37, 45, 48, 49, 51, 52, 54, 104
distribution, 42
district courts, 3, 50
District of Columbia, 37, 67, 96
diversity, 4, 106
domestic laws, 79
draft, 11, 68, 83, 127
drought, 2, 3

E

ecosystem, 47
EIS, 11, 32, 36, 43, 96, 97, 98, 99, 100, 122
election, 65
electricity, 13, 37, 98, 110, 111
emergency, viii, 2, 31, 32, 111
emergency response, 31
emission, ix, 6, 8, 11, 12, 13, 37, 45, 47, 49, 54, 55, 58, 60, 65, 66, 67, 68, 73, 75, 79, 83, 85, 88, 94, 98, 100, 107, 109, 116, 124, 127
emitters, 52, 79, 103, 115
endangered, 91, 92, 120, 121
endangered species, 91, 120, 121
energy, ix, xi, 16, 46, 55, 78, 79, 98, 105, 110, 111, 117, 127
energy conservation, 117
Energy Independence and Security Act, 88
Energy Policy and Conservation Act, xi, 12, 68, 78, 79, 87, 94, 99, 106, 108, 115
enforcement, 9, 60, 122, 124
England, 69
environment, 11, 17, 32, 33, 38, 43, 60, 75, 80, 101, 110, 111, 113, 125
environmental effects, 36, 122

environmental impact, 10, 11, 32, 43, 68, 75, 95, 96, 99, 100, 101
equipment, 89
ERA, 125
erosion, ix, 4, 16, 20, 21, 25, 26, 39, 46, 55, 105
ethanol, 13
Europe, 116
evidence, 54, 69, 99, 110
evolution, 80
exclusion, 16, 112
executive branch, x, 74, 77, 78, 87, 102, 104
exercise, 13, 40, 49, 51, 52, 53, 57, 85, 87, 88
expertise, 54
exposure, 12, 27, 116
extraction, 98
extraordinary conditions, 66, 124
extreme poverty, 33
extreme precipitation, viii, 2, 24
extreme weather events, 3, 16

F

factories, xi, 77
farmers, 102
fear, 33
federal agency, x, 10, 32, 63, 91, 93, 96
federal courts, 47, 49, 104
Federal Emergency Management Agency, 42
federal government, xi, 13, 30, 31, 57, 58, 78, 95, 105, 107
federal law, 49, 87, 106
Federal Register, 65
federal regulations, 49
Fifth Amendment, viii, 2, 10, 19, 21, 22, 23, 24, 27, 40
fish, 14, 19, 20, 57, 113, 121
Fish and Wildlife Service, 35, 91, 99, 120, 121
fisheries, 39, 113
fishing, 90, 91
flexibility, 18, 121

flooding, 16, 23, 24, 26, 29, 30, 40, 47
floods, viii, 2, 23, 30, 43
fluidized bed, 84
food, 43
force, xi, 26, 37, 78, 79, 81, 82, 125
foreign affairs, 109
foreign nationals, 33
foreign policy, xi, 12, 37, 52, 78, 107, 109, 115
Fourth Amendment, 22, 40
fraud, 106
freedom, 33
freezing, 16
fuel consumption, 68, 117
funds, 80, 118

G

GAO, 38
General Motors, 56, 100, 105, 120
Georgia, 34, 120
Global Change Research Act, xi, 78, 101
global climate change, 14, 92
global scale, 114
global warming, 9, 34, 55, 56, 92, 96, 97, 105, 106, 120, 121, 126
God, 126
governor, 108
grants, 65, 66, 67
greed, 110
greenhouse, vii, ix, x, xi, 1, 3, 9, 36, 45, 47, 63, 64, 77, 79, 87, 89, 95, 98, 99, 107, 110, 121, 127
greenhouse gas (GHG), 1, 3, 9, 36, 45, 47, 63, 64, 77, 79, 87, 89, 95, 98, 99, 107, 110, 121, 127
greenhouse gas emissions, 36, 79, 87, 95, 99, 110
groundwater, 19, 39
guardian, 43
guessing, 125
guidance, x, 11, 36, 52, 64, 68, 70, 105, 126
guidelines, 53, 71
Gulf of Mexico, 23, 39, 106

H

habitat, 8, 9, 35, 36, 79, 91, 92, 93, 99, 101, 121
habitats, 91
hardwood forest, 103
Hawaii, 41, 124
hazardous air pollutants, 124
hazardous substances, 31
health, 18, 50, 57, 64, 111, 113
highways, 32
history, 13, 123
homeowners, 16
homes, 27, 102
host, vii, 1
House, 80, 102, 108, 118, 123, 126
human, 3, 11, 18, 39, 47, 75, 102, 104, 113, 114, 125
human health, 47
human rights, 18, 113, 114
Hunter, 123
Hurricane Katrina, ix, 15, 16, 23, 40, 46, 56, 106
hurricanes, 16
hybrid, 12

I

ID, 85
idiosyncratic, 124
immigration, viii, 2
Immigration and Nationality Act, 33
immunity, 37, 40
imports, 37
India, 37
indirect effect, 93, 100
individuals, 17, 113
industry, 69, 73, 79, 108, 112
infrastructure, 30, 31, 42
insurance policy, viii, 2, 111
integrity, 18, 113
interference, 4, 26, 48, 51, 59, 105, 123
international law, viii, xi, 2, 17, 33, 38, 78, 79, 112, 113, 115, 125

investment, 29
Iowa, 58
islands, 31, 41, 42, 115
issues, vii, ix, 1, 3, 4, 5, 11, 15, 16, 19, 24, 30, 36, 45, 48, 49, 51, 52, 55, 66, 68, 70, 85, 87, 96, 109, 117

J

Jordan, 42
jurisdiction, 3, 4, 14, 17, 18, 24, 30, 38, 50, 51, 59, 86, 112, 113, 120, 125

K

kill, 35, 120
Kyoto Protocol, 80, 112, 118

L

landfills, 72
Law of the Sea Convention, 113, 125
law suits, viii, 45
laws, xi, 2, 17, 26, 78, 79, 87, 108, 111, 126
laws and regulations, 126
lead, viii, 2, 13, 25, 30, 31, 83
leakage, 73
legal issues, vii, viii, 1, 2, 3, 5, 11, 25
legislation, 3, 41, 52, 117
levees, viii, 2, 16, 23, 32, 41
liability insurance, xi, 78, 79
light, xi, 5, 6, 7, 47, 48, 53, 58, 68, 69, 73, 78, 82, 94, 99, 107, 111
light trucks, 94
loan guarantees, 98
loans, 98
local government, 29, 31
low-interest loans, 97
lying, 17, 30, 115, 116

M

magnitude, 28, 89, 114
majority, 85, 87, 96, 122
mammal, 92, 120
mammals, 90, 120
man, 16, 47
management, 29, 30, 72, 120, 121
manufacturing, 83
manure, 72
mapping, 101
Marine Mammal Protection Act, 90
mass, 107
materials, 102
matter, iv, 22, 30, 31, 51, 53, 59, 87, 98, 108, 109, 114, 125
melting, 16, 55, 105
membership, 33
Mexico, 58, 98, 110
migrants, 33
migration, 22, 25, 117
minors, 13
missions, vii, ix, x, xi, 1, 3, 45, 47, 66, 77, 79
Mississippi River, 24, 40
Missouri, 34
modifications, 47, 56, 64, 66, 70, 71, 72, 83, 89
Montana, 13, 58, 60, 97

N

National Environmental Policy Act (NEPA), xi, 10, 31, 68, 78, 95
National Research Council, 34, 88
national security, 52
nationality, 33
natural disaster, 33
natural disasters, 33
natural gas, 74, 82, 119
natural resources, 14, 50, 57, 102
Nepal, 114
nitrous oxide, 83, 85, 107
NOAA, 121

nuisance, viii, ix, xi, 4, 15, 29, 34, 45, 46, 47, 48, 49, 51, 52, 54, 55, 56, 57, 58, 60, 76, 78, 79, 90, 102, 103, 104, 105, 106, 110, 123, 124

O

OAS, 18, 113
Obama, 57, 84, 89, 94, 108, 117, 122
Obama Administration, 57, 84, 89, 94, 108
oceans, 10, 97
officials, 40
oil, 15, 55, 56, 58, 82, 90, 95, 97, 98, 99, 105, 106
omission, 97
operations, 72, 91
Organization of American States, 18, 113
Outer Continental Shelf, 79, 95, 97
Outer Continental Shelf Lands Act, 79, 95
overlap, 109
overlay, 41
oversight, 53, 123
ownership, 19, 20, 21, 26, 27, 38

P

Pacific, 35, 90, 92, 99, 115, 116
parallel, 54
PCA, 12, 94, 108
Peru, 114
petroleum, x, 7, 64, 70, 82
pipeline, 12
plants, 30, 53, 70, 83, 91, 97, 111
polar, 9, 36, 90, 92, 93, 94, 97, 99, 120, 121
policy, 4, 5, 16, 17, 40, 51, 52, 53, 56, 85, 87, 88, 95, 104, 105, 106, 111, 119
pollutants, x, 6, 17, 64, 66, 72, 73, 74, 81, 82, 84, 107, 110, 118
polluters, 57

pollution, 5, 6, 8, 14, 17, 34, 49, 53, 57, 60, 64, 65, 67, 74, 81, 85, 104, 106, 112, 113, 116, 120, 124
ponds, 25
population, 19, 27, 124
population growth, 19
power plants, viii, x, xi, 2, 7, 9, 15, 52, 53, 58, 64, 70, 74, 77, 78, 79, 98, 111, 126
precedent, 27, 54, 115
precipitation, 19, 24
preparation, iv, 32, 99, 122
preservation, 18, 20, 39, 113
President, 88, 101, 108, 118, 119, 122
President Obama, 108, 122
prevention, x, 64
principles, viii, 2, 11, 17, 21, 29, 38, 51, 113, 115
private ownership, 26
procedural right, 86, 96, 101, 122
project, 8, 10, 23, 27, 28, 36, 69, 91, 92, 99, 100, 122
property rights, viii, 2, 24, 26, 27, 28, 40, 126
proposed regulations, 88
protection, 18, 28, 40, 41, 56, 106, 111, 113
public health, 6, 8, 48, 53, 64, 67, 80, 81, 82, 85, 87, 88, 103, 120, 123, 124
public interest, 10, 36
public nuisance law, 48, 123
Puerto Rico, 34, 59

Q

quality standards, 8, 81, 120

R

race, 33
rainfall, 30, 102
ratepayers, 110
ratification, 80
reasoning, 53, 90

recall, 27
recognition, 14, 57
recommendations, iv
reconstruction, 28, 32
recreation, 39
Reform, 118, 123
refugee status, 33
refugees, 33, 126
Registry, 116
regulations, 6, 7, 9, 12, 13, 31, 32, 35, 36, 41, 42, 43, 53, 60, 66, 70, 84, 87, 88, 93, 94, 100, 107, 122, 124
regulatory changes, 9
rehabilitation, 43
rejection, 34, 88
relevance, x, 25, 64, 65, 86
reliability, 92
relief, viii, ix, 3, 14, 15, 33, 45, 46, 47, 50, 54, 55, 58, 60, 83, 103, 105, 115, 116, 122
remedial actions, 31
removal actions, 31
repair, 43
requirements, viii, 2, 3, 6, 11, 30, 31, 32, 34, 50, 58, 65, 68, 69, 70, 72, 95, 97, 107, 116, 123
resistance, 23, 116
resolution, 11, 35, 37, 112, 113, 118, 125
resources, 14, 39, 57, 97, 114
response, viii, 2, 31, 87, 89, 111, 117, 118, 122
restoration, 27, 28, 43
restrictions, 10, 26, 29, 30, 41, 80, 93, 100
risk, 16, 24, 37, 100, 127
routines, 7
rules, 7, 21, 22, 37, 51, 68, 73, 90, 120, 121

S

safety, 48, 123
salmon, 92
scarcity, 18, 19

science, 55, 88, 106
scope, 9, 11, 88, 93, 123
sea level, viii, 2, 3, 15, 16, 21, 22, 23, 25, 26, 27, 28, 30, 39, 41, 42, 56, 86, 102, 115, 117
sea-level, 24
sea-level rise, 24
Secretary of Commerce, 36
Securities Act of 1933, 126
sediment, 20
seizure, 22, 40
Senate, 80, 118, 119
Senate approval, 80
settlements, x, 8, 64, 70
shareholders, 126
shifting boundaries, 20
shoot, 35
shoreline, ix, 23, 46
shores, 16
showing, 12, 16, 96, 98
Sierra Club, 43, 83, 84, 119
SIP, 70, 71, 72, 75, 76
social group, 33
solid waste, 72
solution, 117
South Dakota, 110, 124
sovereign state, 86
sovereignty, 4
species, 9, 35, 91, 92, 93, 99, 120, 121
spending, 27
stabilization, 43
standing test, 86, 116
state control, 65
state laws, 12, 60, 100
statutes, xi, 3, 11, 12, 22, 25, 30, 31, 32, 41, 77, 78, 79, 109, 110
statutory authority, 35
statutory provisions, 115
stock, 19, 120
storage, 11
storms, 27, 28, 43, 55, 105
structure, 25, 26, 29, 32, 43
sulfur, 100
Superfund, 31

T

target, 107
Task Force, 118
techniques, 28
technology, x, 6, 11, 64, 65, 70, 74, 83, 89, 119
temperature, 113
Tennessee Valley Authority, 123
territory, 17, 86
threats, ix, 46
time frame, 120
Title V, x, 6, 7, 58, 64, 65, 68, 69, 70, 71, 72, 74
tobacco, 123
tracks, 54
trade, 13, 85, 109, 117
transactions, 126
transmission, 98
transportation, 32, 100
treaties, 80, 112, 113
treatment, 30, 65
trial, 39, 103
triggers, 6, 67
Tuvalu, 116, 126
twist, 24

U

U.S. Army Corps of Engineers, 105
U.S. Department of the Interior, 121
U.S. economy, 118
U.S. immigration law, 33
UN, 43, 113
UNESCO, 114
uninsured, 56, 106
United Kingdom, 116, 126
United Nations (UN), 33, 43, 118, 125
United Nations Framework Convention on Climate Change, 118
United Nations High Commissioner for Refugees, 33

universe, 16
updating, 23

V

vegetation, 22, 126
vehicles, vii, x, xi, 1, 5, 6, 12, 47, 56, 58, 63, 64, 65, 66, 67, 68, 69, 73, 78, 79, 81, 85, 86, 101, 105, 107, 109, 122
vessels, 89
violence, 33
vote, 34, 50, 117
vulnerability, 57, 124

W

waiver, x, xi, 12, 40, 64, 65, 67, 78, 107, 108, 109, 124
Washington, 38, 58, 76, 102, 110, 122
waste, 38, 39, 84
wastewater, 30
water, viii, 2, 16, 18, 19, 20, 21, 24, 25, 26, 27, 29, 38, 39, 41, 56, 59, 60, 92
water rights, viii, 2, 18, 19, 20, 38
water supplies, 19
waterways, 14, 39, 57
welfare, 6, 8, 53, 64, 67, 80, 81, 82, 85, 87, 88, 120, 124
well-being, 50, 103
wetlands, 10, 21, 25
White House, 69
wildfire, 101
wildlife, xi, 14, 35, 57, 77, 79, 91, 101, 121
withdrawal, 38, 94
wood, 119
worldwide, 47, 86
worry, 37

Y

yield, 117